THE
BEAUTIFUL
BRAIN

THE BEAUTIFUL BRAIN

The Drawings of
Santiago Ramón y Cajal

Edited with commentaries by Eric A. Newman, Alfonso Araque, and Janet M. Dubinsky
Essays by Larry W. Swanson, Lyndel King, and Eric Himmel

ABRAMS, NEW YORK

CONTENTS

8 The Beautiful Brain
  Eric A. Newman, Alfonso Araque,
  and Janet M. Dubinsky

11 Santiago Ramón y Cajal
  Larry W. Swanson

21 Drawing the Beautiful Brain
  Lyndel King and Eric Himmel

The Drawings
  33 Cells of the Brain
  83 Sensory Systems
  121 Neuronal Pathways
  157 Development and Pathology

193 Seeing the Beautiful Brain Today
  Janet M. Dubinsky

202 Notes

204 Index

207 Acknowledgments

Frontispiece. Four self-portraits taken by Cajal when he was thirty-four years old, 1886

Previous spread. Self-portrait, taken by Cajal when he was in his late fifties, c. 1910

Left. Cajal (center) posing with family members and friends in Valencia, 1885. Cajal's wife, Doña Silveria, is standing over his left shoulder. The men are members of the Gaster-Club, a social group in Valencia that got together for Sunday excursions, to hike, take photographs, and, as Cajal remembered happily, "enjoy the tasty and famous Valencian paella."

Still life with microscope and laboratory chemicals, photograph by Cajal

THE BEAUTIFUL BRAIN

Eric A. Newman, Alfonso Araque, and Janet M. Dubinsky

Santiago Ramón y Cajal has rightly been credited as the father of modern neuroscience, the study of the structure and function of the brain. Cajal, who lived from 1852 to 1934, was a neuroanatomist who, over the course of five decades, produced more than twenty-nine hundred drawings that reveal the nervous system as we know it today. He studied many aspects of the brain, from the structure of individual neurons (the nerve cells that comprise the brain) and the connections between them, to the changes that occur in the brain during early life and following injury. He did this by examining thin slices of the brain under a microscope. He treated these slices with chemical stains to highlight different types of brain cells and structures within these cells. Most notably, he used a stain developed by the Italian biologist Camillo Golgi, which colors brain cells a deep, rich black. Cajal improved upon the original formulation of the Golgi stain to obtain exquisite images of neurons. Many of the drawings reproduced in this book are based on Golgi-stained brain slices.

Cajal was able to envision the living brain in these dead tissues. It is relatively easy to deduce how the heart pumps blood based on its structure. It is far harder to discern how the brain works from the organization and interconnections of its billions of cells. But Cajal did just that. It has been known since the time of Luigi Galvani in the late seventeen hundreds that information is transmitted within the brain by electrical impulses. However, it was not until a century later that Cajal, in his Theory of Dynamic Polarization, described how information, in the form of electrical signals, travels within individual neurons, from their dendrites to their cell bodies and finally to their axons. Later work proved Cajal to be correct.

In a second fundamental observation called the Neuron Doctrine, Cajal demonstrated that the brain is composed of discrete cells—neurons—rather than a continuous, interconnected network of cell appendages, as most of his contemporaries believed. He also discovered many of the important components of brain neurons, including the dendritic spine, the neuronal appendage that receives signals from other neurons, and the growth cone, the appendage that enables neurons to make precise synaptic contacts with other neurons.

Remarkably, Cajal's detailed studies of the brain are as relevant today as they were a century ago. Cajal's drawings continue to be used because they have never been equaled in their clarity and their ability to express universal concepts. A single Cajal drawing often summarizes a basic principle or a sequence of events much more clearly than could be shown in dozens of photographs. It is not uncommon to see one of Cajal's classic drawings presented at the beginning of a scientific lecture or in a publication, simply because there is no better way of introducing a topic to an audience. The impact of his drawings is due, in no small part, to their intrinsic beauty, which inspires our imagination.

This book presents eighty of Cajal's original drawings of the brain. (A few are of other tissues, such as an insect's leg muscle or a migrating blood cell, illustrating Cajal's wide-ranging interests.) Some of these drawings are well known, while others have not been published previously except in Cajal's original scientific papers. Captions accompanying the drawings describe their subject matter and their scientific importance. Two essays focus respectively on Cajal's life and scientific achievements, and his mastery of the art of drawings. A third essay brings us up-to-date, describing modern neuroscience imaging methods that Cajal, undoubtedly, would have appreciated. We hope you enjoy Cajal's vision of the beautiful brain.

SANTIAGO RAMÓN Y CAJAL

Larry W. Swanson

As long as our brain is a mystery, the universe, the reflection of the structure of the brain, will also be a mystery.
—Santiago Ramón y Cajal

Some nineteenth-century giants of life sciences research are still widely recognized by the general public around the world. The Englishman Charles Darwin revolutionized our thinking about life on earth with his theory of evolution by natural selection, and the Frenchman Louis Pasteur saved countless lives by clarifying the role of microbes in human disease.

Santiago Ramón y Cajal was their equal in scientific achievement—he more than anyone is responsible for creating the modern field of neuroscience—and yet he never attained the fame of a Darwin or Pasteur outside his native Spain or the narrow confines of his scientific profession. Why? Mainly because there is no simple means to encapsulate how Cajal and his contemporaries explained and illustrated the workings of the brain as a biological network in an entirely new way, a way that remains foundational to neuroscientists today.

But Cajal was a fascinating, multidimensional, larger-than-life character, and his main insights into how the brain works are not difficult to understand when following the main threads of his research career, which peaked in 1906 when he shared the Nobel Prize in Physiology or Medicine with the Italian histologist Camillo Golgi. This award has one of the most interesting stories in the history of the prize, because the two scientists held diametrically opposed theories about fundamental brain organization—the basic plan of nervous system architecture. From the beginning, Cajal's view was favored by most scientists,

Self-portrait, taken by Cajal when he was sixty-eight, 1920

Self-portrait, taken by Cajal in his library when he was in his thirties

setting up an intense competition between rival camps of neuroscientists. As we shall see, definitive proof did not come for another fifty years.

Most of what we know about Cajal's early life and scientific career comes from his autobiography, *Recollections of My Life*, often considered among the best scientific autobiographies ever written, perhaps surpassed only by Darwin's *Autobiography*. Cajal was born in Petilla de Aragón, a tiny, impoverished village in northeastern Spain. His father was the son of farmers, and he worked hard, eventually to become the respected local doctor. As a child, Cajal was hardly the studious type, describing himself as shy, unsociable, secretive, brusque, and having an innate dislike for the principle of authority, an utter incapacity for adulation of the powerful. Instead, young Cajal and his friends were often getting into trouble for minor pranks, and in fact his first literary effort, written for friends when he was around fourteen, was called *Estrategia lapidaria*—a treatise on how to design and use slingshots. From an early age, he showed an obsessive-compulsive streak that he referred to as "manias." Drawing everything in sight at age eight (his self-described graphomania) was followed by collecting as many kinds of birds, birds' eggs, and birds' nests as possible, building homemade "cannons," bodybuilding, playing as many simultaneous games of chess as he could, and on and on.

But two of his preoccupations—drawing and photography—were to play major roles in his future career as a scientist. Of course, father wanted son to follow in his footsteps, to become a physician, but this was furthest from Cajal's mind at the time. Instead, he wanted to be an artist. Trying gently to put a stop to this ambition, Cajal's father persuaded an itinerant plasterer-decorator to assess his son's artistic talent—which as expected was judged basically nonexistent. This episode had no detectable effect on Cajal's artistic production, however, and a few years later, at age sixteen, he was inspired by Daguerre's invention of practical photography and taught himself the art—taking, developing, and printing the first of a series of brilliant self-portraits that documented almost every stage of his life.

Cajal's father finally saw an opening for collaboration—persuading his son to help him teach human anatomy to students at the nearby medical school in Zaragoza, and this venture proved quite successful. Cajal enjoyed learning about the human body and used his talents to produce excellent drawings of anatomical dissections intended for an anatomical atlas—a tradition going back to the artists in Titian's studio, who illustrated the monumental *De corporis humani fabrica libri septem* of Andreas Vesalius, published in Basel in 1543.

Inspired, Cajal enrolled at the medical school in Zaragoza and graduated with a licentiate in medicine at age twenty-one, in 1873. He was sent that year by the army medical service to Cuba, where he contracted malaria and as a result was discharged after about a year of service. He recovered, but the disease left him with a weakened constitution, so much so that the more contemplative life of a medical school professor was chosen over the more physically demanding life of a practicing physician like his father. Cajal basically started at the bottom in 1875, with a temporary assistantship in anatomy at Zaragoza, where out of his own pocket he equipped a home laboratory and began learning the methods of histology—the microscopic examination of body tissues—as opposed to the human gross anatomy he learned with his father.

Human anatomy has been taught regularly in medical schools since the time of Vesalius because it is an excellent way to learn how the body works and to discover the cause of death at autopsy, and it is, of course, the starting point for surgery. Histology, on the other hand, really only began to flourish in the 1830s with the development of powerful microscopes by the German optical industry. By the end of the decade, the conceptual framework of histology, the cell theory, had been laid in the classic publications of Schleiden (1838) and Schwann (1839). Remarkably, the cell theory described the first new level of biological organization since the foundational research of Aristotle in classical antiquity on organs and tissues. Basically, the cell theory proposed that all tissues in animals as well as plants are made up of individual microscopic units called cells, with different tissues having different arrays of cell types. The nature of cells making up the nervous system (brain, spinal cord, and nerves) has been the most difficult problem in histology from the beginning.

Cajal's first scientific publication was on histological changes seen during the inflammatory response to injury, a topic of obvious medical importance. It was published in 1880 and was based on research carried out in his personal laboratory at home. That year he also married Silveria Fañanás García, who eventually assumed major responsibility for raising their four sons and four daughters. Three years later, thirty-one-year-old Cajal assumed the chair of anatomy at the larger Faculty of Medicine in Valencia, where he immediately began work on an original textbook of histology, the first of its kind in Spanish, that was published in 1889.

At the time, microbiology, exemplified by the work of Pasteur and many others, was by far the most exciting, popular field in medical research, but Cajal deliberately chose a different route instead, writing, "I finally chose the cautious path of histology, the way of tranquil enjoyments. I knew well that I should never be able to drive through such a narrow path [microbiology] in a luxurious carriage; but I should feel myself happy in contemplating the captivating spectacle of minute life in my forgotten corner and listening, entranced, from the ocular of the microscope, to the hum of the restless beehive which we all have within us."[1]

From his teaching and textbook writing, Cajal was acutely aware of the major unanswered question in nineteenth-century neurology: How is the nerve impulse transferred from one nerve cell to another in the adult nervous system? For example, what is the relationship between a sensory nerve cell that detects painful stimuli in the skin and a motor nerve cell in the spinal cord that moves a muscle to escape the painful stimuli? In other words, what is the cellular basis of reflexes, a basic protective function of the nervous system that is critical for health and survival?

In a flash of insight, Cajal saw a way to answer the big question on a visit to Madrid in 1887. He was visiting a friend, Luis Simarro Lacabra, who had just returned from Paris, where he learned the most recent advances in histological technique. The fateful moment came when Cajal peered down the microscope at a slide with a bit of nervous tissue prepared with the Golgi method. Other histological methods showed all of the nerve cells, packed together in an inextricable tangle. But the method discovered accidentally by Golgi in 1873 was different. Only occasional nerve cells were observed, but they were stained in their entirety and in a beautifully clear way: deep black silhouettes against a light yellow background that lent themselves readily to pen-and-ink drawings. It was a seemingly miraculous result, and one whose mechanism of action is still not fully understood today.

Left. Silveria Ramón y Cajal, photograph by Cajal

Right. Cajal with four of his children, (from left) Fe, Jorge, Paula, and Santiago, in a photograph that he took when he was thirty-six

Curiously, the method had been used in the preceding fourteen years only by Golgi and a small group of his students and visitors to his laboratory in Pavia. Returning to Valencia, Cajal quickly learned why: Golgi's method was both time-consuming and very capricious—different batches could yield very different results. So Cajal worked tirelessly to improve the method's reliability and usefulness, and the results were spectacular. He wrote that 1888 was "my greatest year, my year of fortune. . . . the new truth, laboriously sought and so elusive during two years of vain efforts, rose up suddenly in my mind like a revelation. The laws governing the morphology and connections of nerve cells in the gray matter, which became patent first in my studies of the cerebellum, were confirmed in all the organs which I successively explored."[2] He continues, "Realizing that I had discovered a rich field, I proceeded to take advantage of it, dedicating myself to work, no longer merely with earnestness, but with a fury."[3] He often worked fifteen hours a day, complaining that each discovery cost a night's sleep, and in 1890 alone published an astounding fourteen scientific articles on the nervous system.

What laws of the nervous system did Cajal discover in this frenzy of scientific discovery? The answer lies in a comparison with the interpretations Golgi himself provided for

the relationship between nerve cells. Golgi showed very clearly with his method that nerve cells have two fundamentally different types of long, thin extensions from the cell body, the main part of the cell that also contains the nucleus and its chromosomes.

Golgi's observations led him to speculate that one type of nerve cell extension, now referred to collectively as dendrites, is essentially nutritive in function, like the roots of a tree, whereas the other type of extension, now referred to collectively as axons, transmits neural impulses between nerve cells, and between nerve cells and other cells (like muscle or gland cells). Thus, for Golgi, connections between nerve cells are formed exclusively by axons. But more than this, he thought that nerve cells are directly connected to one another by their axons—so that nerve cells form an uninterrupted network in an arrangement called a reticulum, like a spiderweb; this was the Reticular Theory of nervous system organization, and it was in vogue when Golgi made his observations. His descriptions were crafted in light of this prevailing theory.[4]

Also using Golgi's method, Cajal came to radically different conclusions, which ultimately proved correct and showed that Cajal was the better observer, and a more original, critical, and insightful thinker. Basically, Cajal provided the conceptual framework for

Cajal was fascinated by the medium of photography, and in addition to his self-portraits, family portraits, still lifes, and microphotographs, he enjoyed taking a stereoscopic camera on excursions and trips. Here are his photographs of young women on the beach at Biarritz, France (above), and street entertainers in Madrid (opposite).

thinking about the cellular wiring diagram of the brain and nervous system that is still in use today. This framework rests on two fundamental principles or laws, and they were first discovered in the cerebellum of the bird and then systematically confirmed in virtually all other parts of the nervous system.[5]

The first principle has long been known as the Neuron Doctrine, and it is simple to state: The nerve cell (neuron) is the structural and functional unit of nervous system circuitry, and nerve cells interact with other nerve cells by way of contact or contiguity—they are not generally in direct continuity as stated by the Reticular Theory. The exact nature of interactions between nerve cells, and between nerve cells and other cells, was not really understood until much later, in the 1950s, when the electron microscope was able to visualize a very narrow gap (cleft) between nerve cells that forms part of the communication unit called a synapse.

Cajal called the second principle the Theory of Dynamic Polarization. Functional polarity, as it is more often described, states that during normal operation, information flows through a nerve cell, and thus through neural circuits, in one direction: from dendrites and cell body to axon. In other words, dendrites do not function like roots (as Golgi proposed); they act as the input side of nerve cells. And conversely, axons are the output side of nerve cells, and they are not in direct continuity with one another as Golgi also proposed. As Cajal first observed in the bird cerebellum, axons and their collaterals typically end in a swelling (the axon terminal), and this swelling lies adjacent to a dendrite or cell body. We now know that electrical neural impulses travel down an axon to its terminal, where a chemical neurotransmitter is released into the synaptic cleft to act on the postsynaptic dendrite or cell body.

The definitive summary of this research was provided in a great classic of modern neuroscience, the two-volume *Histologie du système nerveux de l'homme et des vertébrés* (1909, 1911), which contains more than a thousand illustrations by Cajal. Monumental as this body of work was, Cajal then went on to explore a whole new field of experimental neurology, his findings from which were published as another two-volume masterpiece, *Estudios sobre la degeneración y regeneración del sistema nervioso* (1913, 1914). But there was a whole other side of the man. He wrote one of the first books on the technology and art of color photography, and he remains a major literary figure in Spain. In addition to the wonderful autobiography mentioned at the beginning, he wrote an entertaining book of aphorisms, *Charlas de café*, that remains in print today; *Advice for a Young Investigator,* a book of fatherly advice to young scientists; *El mundo visto a los ochenta años,* a book on the observations of an arteriosclerotic octogenarian; and even a collection of science-fiction short stories.

The great man died peacefully on October 17, 1934, and his funeral was a simple one. It was attended by huge crowds of Spaniards from all walks of life, and he was buried in the necropolis of Madrid next to his wife of fifty-four years.

Still life with oranges, grapes, a rose, a geranium, and a bouquet, color photograph by Cajal, using a technique that he invented, c. 1912

DRAWING THE BEAUTIFUL BRAIN

Lyndel King and Eric Himmel

The human form is Western art's greatest subject. However, anatomical illustration has played a small role in art's history. The Greeks and Romans had a deep understanding of surface anatomy—what can be seen without dissection—as evidenced by their accurate depictions of the human body in sculpture, but this knowledge was lost during the Middle Ages, when art was viewed as the chief ally of the spiritual world. In the fifteenth century, especially in Italy, when the focus shifted back to humanity rather than God as the center of the universe, artists began to take a renewed interest in surface anatomy as a way to more accurately depict the nude figure, while physicians brought artists into the dissecting room for the first time to make a visual record of their investigations. Many magnificent books of anatomical illustrations were produced over the succeeding centuries, but only one artist emerged whose drawings of the interior of the human body appealed to the wider audience of people who loved art. That was Leonardo da Vinci, who was driven by curiosity to extremes of scientific investigation. Lost for four hundred years, Leonardo's anatomical drawings are revered today not only as scientific illustration, but also for their expression of a new concept of human nature emerging from living flesh.

    The same can be said of the Spanish neuroscientist Santiago Ramón y Cajal, whose artwork has been known up until now mostly through illustrations in scientific publications, and who, like Leonardo, was preternaturally gifted in art and science. Cajal's humane vision of a complex system of information-processing circuits emerging out of living, changing tissue is no less emblematic of our time than the curiosity that drove Leonardo to open up the human body and draw what he found is of the Renaissance. His legacy of almost

Cajal posing with a friend, taken by Cajal in his late teens, c. 1870. This was possibly a study for a painting.

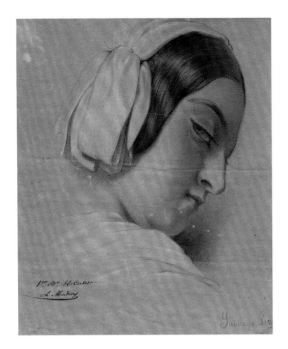 

three thousand drawings that reveal the microscopic anatomy of the brain as he saw it and as we now understand it is unprecedented in the history of modern science.[1] There is no comparable example of such a searching scientific account of basic facts of nature being rendered in such expressive drawings made by its author. Cajal's feelings about art were so intense that he divided his masterful autobiography, *Recollections of My Life*, neatly in half: The first part offers an impassioned and conflicted exploration of the role of art in his life; the second part is a calmer account of his career as a scientist. In fact, science and art were both woven into the fiber of his being.

Cajal recalled the passions of his childhood (in addition to making mischief) as the close observation of nature and drawing. Of his love of nature, he wrote, "All the hours of freedom . . . were spent in wandering about the outskirts of the town exploring glorious ravines, floodplains, springs, rocks, and hills."[2] Of his drawing, he described it as an "irresistible mania for scribbling. . . . Translating my dreams onto paper, with my pencil as a magic wand, I constructed a world according to my own fancy."[3] Cajal bitterly described his father's opposition as a decisive deterrent to a career as a painter, but that didn't prevent him from obtaining some training in the arts. As a teenager at the Academy of Arts in the provincial capital of Huesca in 1866, he drew from plaster casts of ancient Roman and Greek sculpture and copied Renaissance drawings, much as Leonardo's contemporaries learned their craft. But from the youthful work that survives in Cajal's archive, it's clear that he was happier out in the warm Spanish sun making closely observed watercolors of local landscapes than he was in the academic drawing studio, where his portraits came out as stiff and unnatural (above).

When Cajal was about sixteen, his father made him his accomplice in grave robbing to obtain bones for medical anatomical study. The bones, Cajal wrote, were merely one more

Left. Portrait of a young girl, drawing by Cajal for his art class in Huesca when he was sixteen, 1868

Right. Landscape with the Chapel of Our Lady of Casbas, Ayerbe, watercolor by Cajal when he was in his late teens, c. 1871

subject for his pictures, but the drawings that he made of them amazed his father and his teachers. Cajal's nemesis became his master through three intense years of apprenticeship in dissection and anatomical drawing: "My pencil, which was formerly the cause of so much bitterness, at last found grace in the eyes of my father. . . . Gradually, my anatomical watercolors grew into a very large portfolio of which my father was quite proud."[4] By Cajal's own account, pride—both his own and others'—in his artistic gifts would be a constant source of courage as he climbed from the backwaters of Spain to the intimidating heights of international fame.

Cajal followed that steep path to fame as a scientist, not as an artist, but he always maintained that his thought process was based in visual experience and expression: "I am what is called a visual type." As a young man, he learned by seeing, observing, and taking things apart, rather than by reading, memorizing, or listening to dry lectures, which he associated with beatings when he wasn't able to regurgitate information on exams. "My memory was poor for miscellaneous words . . . but such memory weakness was much diminished when the word and the idea were associated with some clear and vigorous visual perception." A typical childhood feat was drawing the map of Europe from memory, "without going astray even in the complicated geography of the German confederation." Later, his heroic exploits of observation became legendary: "Once I spent twenty hours continuously at the microscope watching the movements of a sluggish leukocyte in its laborious efforts to escape from a blood capillary."[5]

Cajal's chosen profession—the study of the microanatomy of the brain—would demand such prodigious feats of visual memory and close observation. To understand why this is so, it helps to have a bit of context. By the time Cajal was choosing a medical specialization in the 1870s, the brain and nervous system were still terra incognita. Most of the body's other major systems—what is called its gross anatomy—are scaled within the threshold of human vision and so were able to be drawn in the dissecting room, without microscopes. In the dissecting room, the brain has shape and structure and its connections by nerves to other organs are visible, but when cut into, it presents a relatively undifferentiated mass of gray and white matter. The key elements that make the brain unique, its multitude of different cell types, cannot be seen with the naked eye. And so explorers of the brain were very much in the same situation as astronomers: Progress could be made only when visualization tools advanced the human ability to see. In the case of the brain, these tools included microscopes (to enlarge small brain slices) and chemical stains (to bring out details in the slices), which were gradually improved through the nineteenth century. And because microphotography was not yet adequate to the task of collecting detailed enough images, anatomists of the brain also needed to be able to draw in order to communicate their findings.

The brain is a great prize, and some of Europe's most talented scientists devoted themselves to discovering its secrets in the late nineteenth century. This was challenging and controversial work: At best, a brain slice seen through a microscope is notoriously difficult to interpret. To borrow one of Cajal's favorite metaphors (see page 41), imagine entering a forest with a hundred billion trees armed only with a sketchbook, looking each day at blurry pieces of a few of those trees entangled with one another, and, after a few years of this, trying to write an illustrated field guide to the forest. You won't get anywhere if you simply draw what you see every day; you're going to have to build up a mental inventory of rules for the forest, and then scrupulously try to fit what you see into that framework, or be flexible enough to allow what you see to reshape your stock of ideas.[6] Among his peers, Cajal was best at this demanding regimen of thinking and drawing, and he emerged from the neuronal forest with the best field guide. As the scientist who was commissioned by the Nobel Committee to weigh the nominees reported in 1906, "it is [Cajal] who has built almost the whole framework of our structure of thinking."[7]

We can picture Cajal at work from one of his best-known photographic self-portraits (opposite), taken in his laboratory in Valencia when he was in his early thirties. His microscope is on the worktable where he is seated, and to its right is his small drawing surface. We may imagine him dividing his attention between the image in the microscope and his drawing paper as he sketched. He preferred to work freehand, rarely resorting to a camera lucida, a device that projects the image from a microscope onto paper where it can be traced. He might start a drawing in pencil, and then later go over it in India ink, adding ink washes or watercolor for tonal areas. Often, he would spend a morning at the microscope without sketching at all, and draw from memory in the afternoon, returning to the microscope to confirm and revise his observations; we can see traces of this process in whited-out areas that he wasn't happy with (see page 64).[8] Finally, because each drawing was intended for publication in a scientific book or article, he inked in reference letters, which would have referred to a key that was set in type under the printed reproduction. Because the blotches of white correcting wax would not have shown up in the published versions of the drawings, nor would the pencil lines have been as evident as they are in some of the original drawings, he didn't bother to disguise them. But he cared a great deal about how the drawings would look in print, and lacking the resources to pay for fine printing early in his career, he even made photolithographic plates for the printer himself.

His drawings were both observations and arguments. If he wanted to draw attention to specific cell types, these might be drawn darkly, while other cells would be drawn more lightly or indicated with a wash (see pages 52 and 124). He would also enlarge cells out of proportion to their surroundings if he wanted to emphasize them (see page 68). Furthermore, we know that, as in the case of an artist who sketches from nature and then creates his painting in the studio, Cajal often combined images taken from different brain slices (see page 56). Sometimes, the composite takes the form of a series of time-lapse images (see page 159). These latter drawings were wholly original compositions, mental reconstructions of a complex reality viewed in discrete parts. It's possible that his most famous drawing, the pyramidal neuron (see page 36), was based on a specific cell in a single brain slice, but it is more likely that he composed it after looking at hundreds of brain slices: It shows evidence of many of the same aesthetic decisions an idealized portrait does.

Self-portrait, taken by Cajal in his laboratory in Valencia when he was in his early thirties, c. 1885

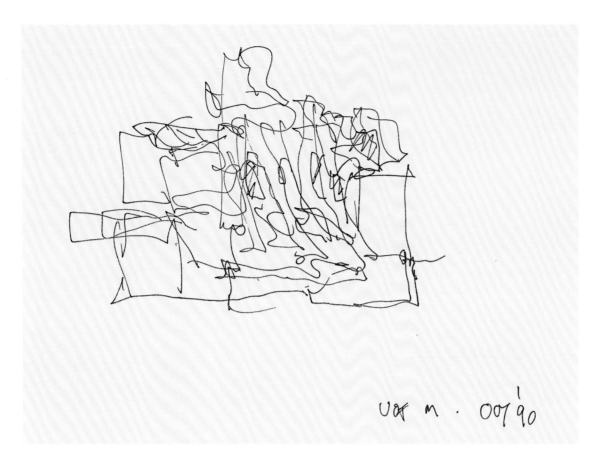

Drawing by Frank Gehry for the Weisman Art Museum, 1990

Milton Glaser, the graphic designer and masterful communicator of visual ideas, describes the power of drawing as an aid to perception: "When I look at something, I do not see it unless I make an internal decision to draw it."[9] "What has not been drawn has not been seen" was a credo of nineteenth-century biology, and Cajal's way of putting this was even more decisive: "A graphic representation of the object observed guarantees the exactness of the observation itself."[10]

But more than a means of honing his vision, drawing was, for Cajal, "a language, an articulation of ideas which allowed thoughts to develop," as intellectual historian Laura Otis puts it.[11] Examples of Cajal's thinking through drawing abound in this book, such as his diagrams showing why information from the eyes travels to opposite hemispheres of the brain (see pages 86–87). But perhaps none are more revealing than his drawings showing contrasting theories of the composition of the brain (see page 146–147), where his picture of the Reticular Theory propounded by his scientific rivals degenerates into a poorly drawn web of lines, as if Cajal's thoughts and his drawing hand were equally repelled by the concept. Here, his hundreds of hours spent looking at and drawing neurons would have reinforced his intellectual insight that a single massive network could not account for the existence of specific brain circuits: "To affirm that everything communicates with everything else is equivalent to declaring the absolute unsearchability of the organ of the soul."[12]

This idea of drawing as a form of thinking is another legacy of Leonardo's art world. In Renaissance Italy, the art of drawing, or *disegno*, referred specifically to the process of thinking through a problem visually, whether that problem was a painting, a sculpture, a building, or a scientific puzzle such as the musculature of the human torso. Good *disegno* combines draftsmanship, or artistry, and invention, or thought. Lyndel King, one of the authors of this essay, experienced the power of *disegno* firsthand, when the University of Minnesota hired architect Frank Gehry to design the new Weisman Art Museum, completed in 1993. Among architects, Gehry is known for loose, exploratory sketches that *are* his ideas for buildings as they're being born: "That's the way I thought out loud."[13] It was thrilling for King to realize, when she looked back at them, that the basic concept of the finished building was there in Gehry's earliest, sketchiest drawings, and all the models and later renderings were about working through the details. The drawing reproduced opposite captures the raw energy of the sculptural forms that later became the building's river façade, relating it to the flow of the Mississippi River in the gorge below. In the end, even though Gehry draws what he sees in his imagination and Cajal what he observed in nature, they are both masters of thinking by drawing. And they both produced astoundingly vibrant drawings—and ideas.

That Cajal's drawings remain living documents a century after they were created is at least partly owing to this vitality, which draws on fantasy and the imagination more than we might expect in a scientific project. Cajal's forms are drawn with clarity, though never mechanically, and his line is confident and constantly moving: Dendrites and axons, the brain's wiring, seem to pulse with life, twisting and turning and bulging and narrowing (see pages 60 and 99). Spending several days under the spell of his conversation in London, the neurobiologist and Nobel laureate Sir Charles Scott Sherrington was struck by Cajal's "intense anthropomorphism," which carried him, "through his microscope, into a world populated by tiny beings actuated by motives and strivings and satisfactions not very remotely different from our own."[14] Perhaps Cajal was winking at Sherrington when he portrayed a damaged Purkinje neuron as a comical swimming penguin (see page 170). Cajal famously used the everyday language of nature to describe the landscape of the brain: "In our parks are there any trees more elegant and luxurious than the Purkinje cells [see page 48] from the cerebellum or the psychic cell [see page 36] that is the famous cerebral pyramid [pyramidal neuron]?" He named dendritic spines with the Spanish word for thorn, *espina*, evoking the canes of a climbing rose, and his drawing of dendrites covered with these "thorns" (see page 44) would not be out of place in a book of botanical art.[15] These images of animals and trees and plants growing in the brain should not be surprising, coming from the boy who first used drawing "to translate my dreams onto paper," and they express a taste for the surreal and folkloric, beyond their all-embracing naturalism. More surreal still was Cajal's fantasy of a man with microscopes for eyes, which appeared in a privately published book of science-fiction stories.[16] Not surprisingly, Cajal did come to the attention of the Surrealists, via Salvador Dalí, Federico García Lorca (see page 29), and Luis Buñuel, who encountered his work in Madrid in the mid-1920s.[17] The Surrealists would have had no inhibitions about cutting open the brain to see if it contained the key to dreams, and many people have wandered into Cajal's lyrical, organic forests and noted visual parallels to Surrealist drawings.

Cajal's work is inescapably modern, notwithstanding that his sense of beauty was formed in the nineteenth century. But Cajal did not view himself as a modernist. In *The World Seen at Eighty*, caustically subtitled *Impressions of an Arteriosclerotic*, he disparaged modern art: "a contradictory jumble of schools that have christened themselves with the pompous names of avant-garde, Cubism, Expressionism, Fauvism, Postimpressionism, etc." He excoriated critics who praised artists for rejecting the "slavish copy of the natural . . . as if the strict copy of nature . . . was unable to communicate feelings and ideas."[18] He could never be comfortable in an art world that had rejected nature as he saw it—a world where Cubists tore nature apart and reassembled it in paintings of jagged forms. In this, Cajal was no different from Albert Einstein, whose taste in the arts was notoriously conservative, in spite of his having conjured a view of nature that was far stranger than anything the Cubists could have imagined.

As with Einstein's theories, it has taken many decades for those of us who are not scientists to catch up with Cajal's brain. It wasn't until 1946, twelve years after Cajal's death, that the first electronic computer flickered into life, hinting that a machine could be built that behaved like a brain. That day may be (infinitely) far off, but since then, the concepts that Cajal discovered, explained, and illustrated have burrowed into the world's technology, economy, popular myths, moral dilemmas, philosophical debates, and art and literature. Since Cajal, we have seen mounting evidence that the idea of the brain being as vast and mysterious as the universe—for centuries a trope for poets—may contain some literal truth. When we look at his drawings today, we see not diagrams or arguments, but the first clear pictures of that remote frontier, drawn by the man who traveled farthest into its endless reaches.

Untitled Surrealist drawing by Federico García Lorca, 1927

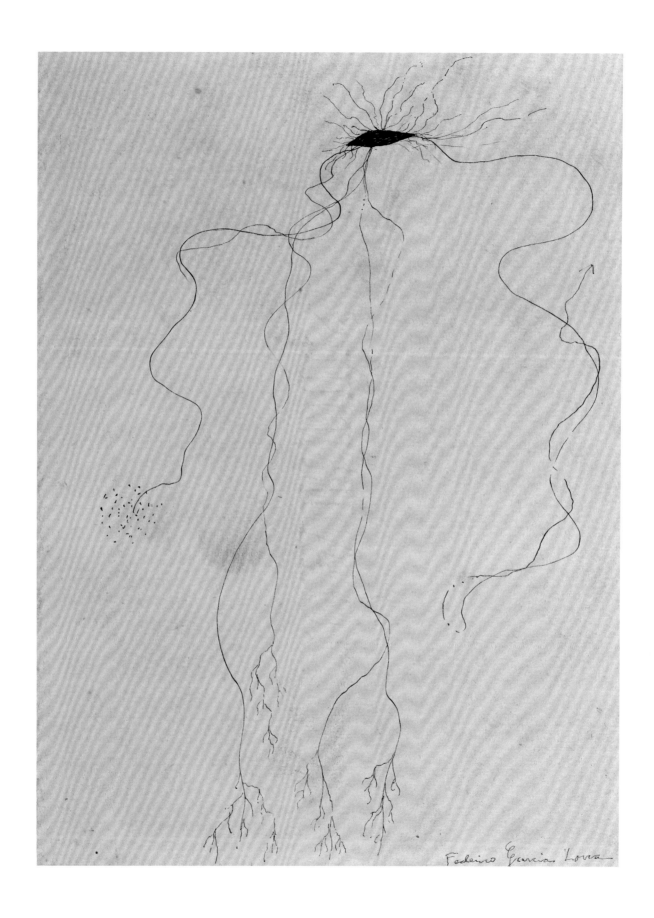

THE **DRAWINGS**

Calyces of Held in the nucleus of the trapezoid body (see page 105)

Fig 10 3/4

C

D

A

B

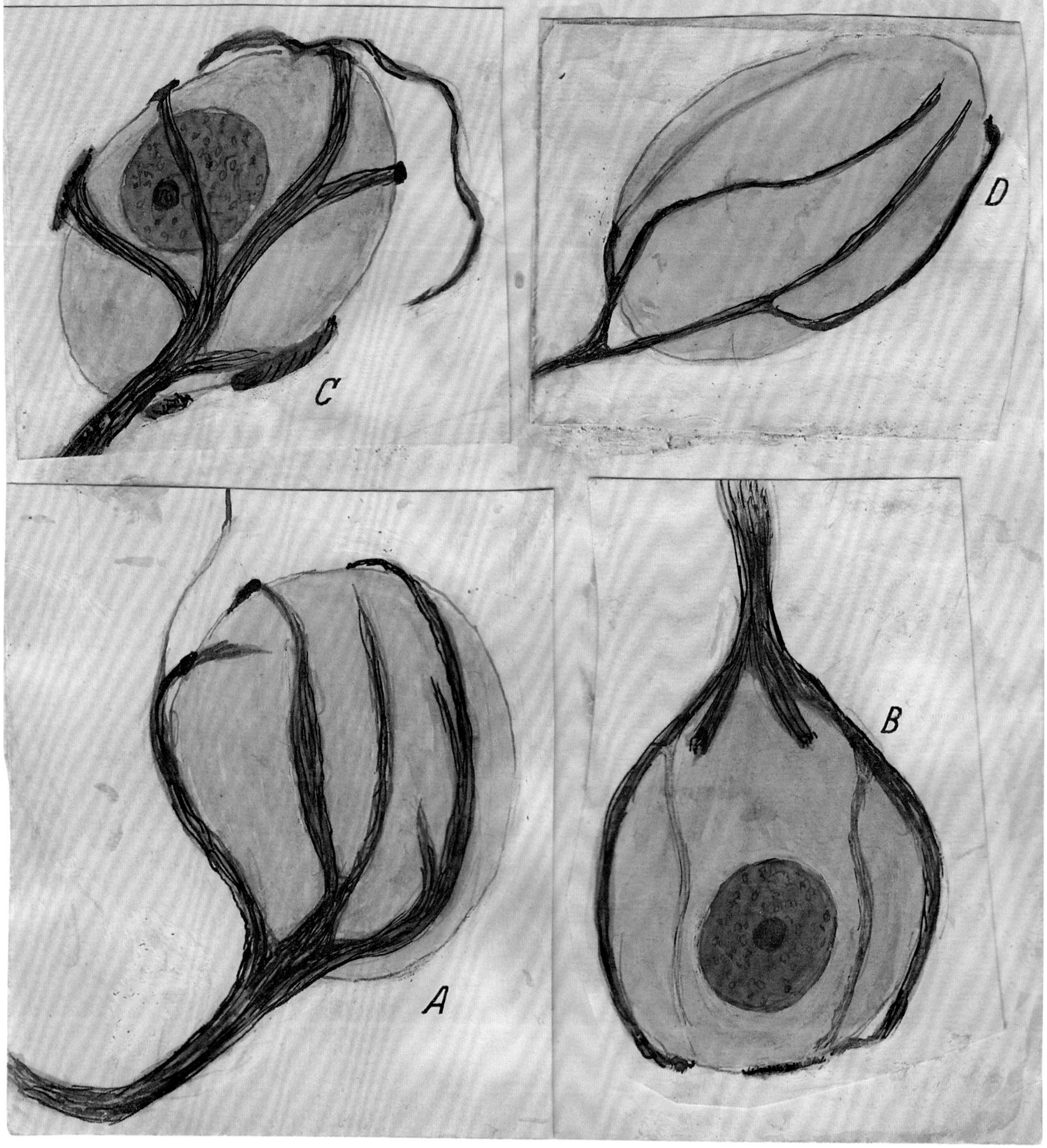

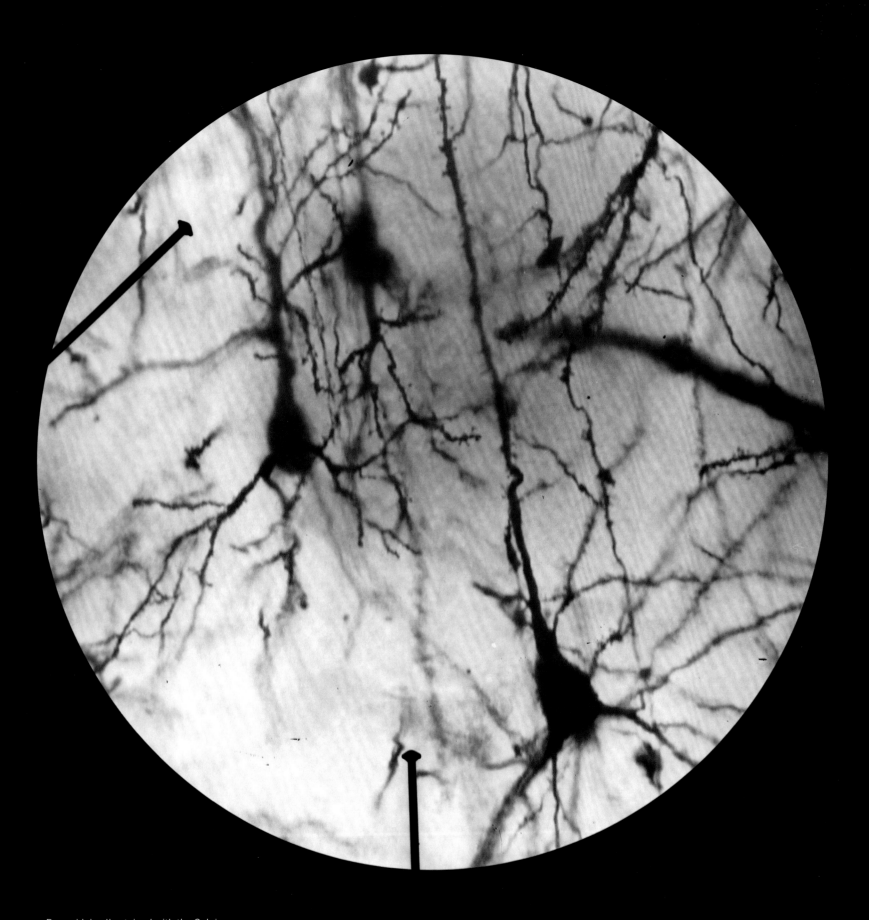

Pyramidal cells stained with the Golgi method, microphotograph by Cajal, 1918

CELLS OF THE BRAIN

Cajal's greatest contribution to our understanding of the brain was, arguably, his championing of the Neuron Doctrine, which held that the brain is composed of discrete cells (neurons) rather than a continuous network of cellular appendages. Early in Cajal's career, this latter theory, the Reticular Theory of the brain, held prominence. Microscope observations of the brain led scientists in the mid- and late nineteenth century to believe that, unlike other organs in the body, which were known to be composed of discrete cells, the brain was made up of one large network of continuously interconnected cells. Cajal, using his keen observational skills and newly developed staining techniques that allowed him to see brain cells in great detail, realized that the continuous network that others had seen was actually made up of discrete cells that were separated by gaps. Cajal won over many, but not all, scientists of his time to the Neuron Doctrine. With the advent of the electron microscope in the 1950s, which magnified images of the brain to a much greater extent than the light microscopes used by Cajal and his colleagues, the Neuron Doctrine was conclusively confirmed. Electron microscope pictures clearly show that individual neurons are separated by gaps one-millionth of a millimeter wide. These gaps between neurons are the synaptic contacts where signals are transmitted from one neuron to another by the release of neurotransmitter chemicals.

Cajal and his contemporaries recognized that, in addition to neurons, the brain was composed of a second class of cells, the glial cells. Neurons and glial cells differed greatly in shape. Neurons possessed a dendritic tree that sprouted from the cell body. These trees received the synaptic inputs from other neurons. Neurons also possessed an axon that emerged from the cell body and could travel to distant parts of the brain. Cajal knew that neurons generated electrical signals and were responsible for the main functions of the brain, receiving sensory inputs, processing this information, storing memories, learning, and controlling our muscles. Glial cells, on the other hand, lacked dendrites and axons and did not generate electrical signals.

The most common type of glial cell in the brain, and the one that Cajal studied in greatest detail, was the astrocyte. Cajal developed specialized methods for staining astrocytes so that all of their fine morphological details were visible. Astrocytes possessed multiple appendages that branched out from their cell bodies. Some of these appendages surrounded adjacent neurons. Others, called endfeet, contacted blood vessels. During Cajal's lifetime the functions of glial cells were uncertain. Speculation about glial cell functions included providing structural support and nourishment for neurons, insulating neurons from each other (a theory proposed by Cajal), and even participating in the information processing functions of the brain. Based on the contacts between glial cells and blood vessels, Cajal suggested that glial cells might regulate blood flow in the brain. To this day, the functions of glial cells in the brain remain controversial. However, we now know that all of these early speculations were, at least to some degree, correct.

The drawings in this section illustrate the neurons and glial cells that Cajal observed and characterized in great detail. Most notably, this includes the pyramidal neuron, which Cajal referred to as "the noble and enigmatic cell of thought"; the Purkinje neuron with its large, elaborately branched dendritic tree; and the astrocyte, the glial cell that forms intimate connections with neurons.

THE NEURON: As Cajal demonstrated more than a century ago, the brain is composed of discrete nerve cells called neurons. The neuron is composed of several important parts. The dendrites, a series of branched appendages, receive inputs from other neurons. Signals are received at synapses, the contacts between neurons where chemicals (neurotransmitters) convey the signals from one neuron to another across a small gap. In many neurons, synapses occur on fine, hairlike extensions of the dendrite called dendritic spines. Once received by the dendrites, signals are transmitted to the cell body of the neuron and then to the axon, a thin appendage that conducts the signals away from the cell body. The axon ends in additional synapses, where contacts onto other neurons are formed. The axon can be short, less than a tenth of a millimeter, or very long, greater than a meter.

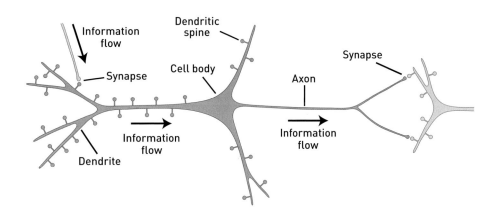

THE BRAIN: Cajal described the neurons and the connections between neurons in almost every part of the brain. This diagram illustrates some of the structures in the human brain that are represented in this book. The structures are shown as they appear at the midline of the brain. (Although the two hippocampi are not present at the midline, as they lie off to the side, their position is also indicated.) The cerebral cortex is responsible for our higher brain functions and contains pyramidal neurons (pages 36–47 and 172–173), which Cajal described in detail. The hippocampus consolidates our memories and also contains pyramidal neurons (pages 136–143). The cerebellum, which facilitates fine control of our movements, contains Purkinje neurons (pages 48–51 and 164–171). The thalamus, a relay station for sensory information from our sense organs to our cortex, lies deep within the brain (pages 114–117). The brainstem helps to control our breathing, heart rate, and movements, among other functions (pages 31, 104, 108, 145, 152). The spinal cord, an integral part of the central nervous system, serves as a conduit for signaling between our brain and our body and coordinates our movements (pages 76, 146–151).

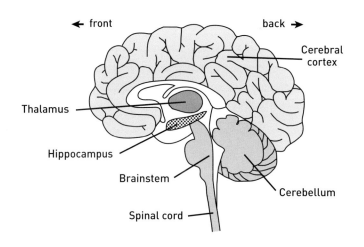

CELLS OF THE BRAIN

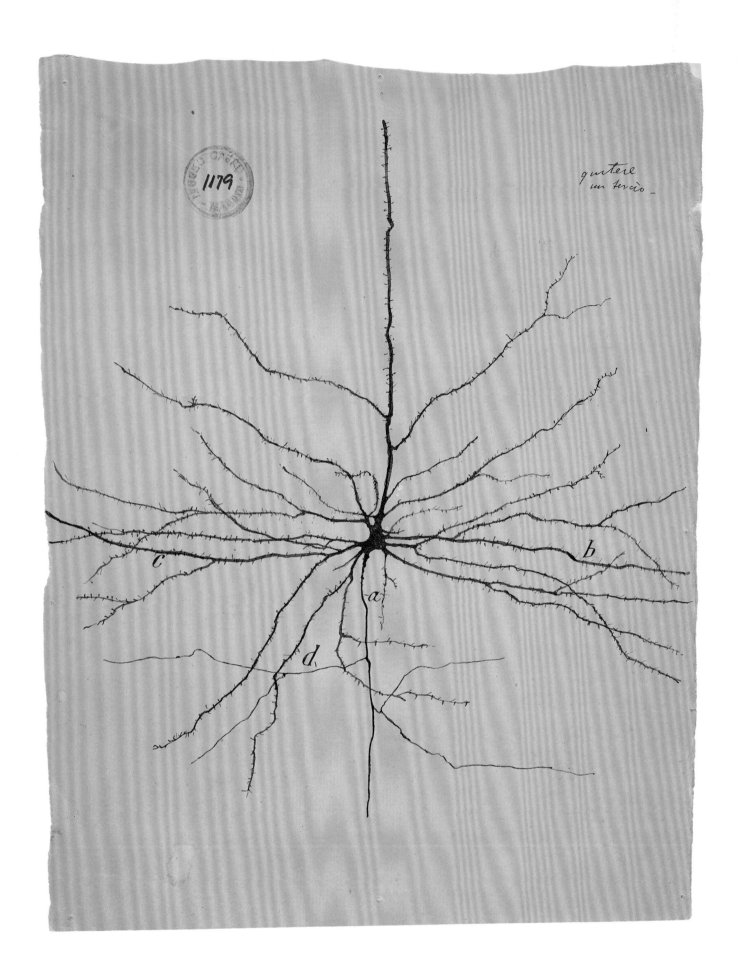

The pyramidal neuron of the cerebral cortex

The cerebral cortex is the outermost layer of our brain. It receives and processes information from our sense organs, commands motor activity, and is responsible for higher brain functions. Pyramidal neurons, which are critical to the function of the cerebral cortex, were characterized in great detail by Cajal. The pyramidal neuron derives its name from its pyramid-shaped cell body, the large structure at the center of the cell. These neurons are found both in the neocortex, the evolutionarily advanced portion of the cerebral cortex that is involved in higher brain functions, and in the hippocampus, a more primitive portion of the cortex where many of our memories are first laid down. Like many neurons in the brain, the pyramidal neuron is axially symmetric around the large dendrite emerging from its cell body. This symmetry is beautifully conveyed in this drawing, where Cajal artfully varies the weight of the dendrites to give the viewer a sense of the three-dimensionality of the dendritic tree. Because of their large size, pyramidal neurons are among the few neurons in the brain that can be seen with the naked eye, without the benefit of a microscope. A single pyramidal neuron is illustrated in this iconic image.

A giant pyramidal neuron of the human cerebral cortex

Cajal observed that pyramidal neurons come in different sizes and shapes. The pyramidal neuron illustrated here is a giant one. Its cell body lies deep below the surface of the cerebral cortex. Branching upward from its cell body is one set of very long dendrites over a millimeter in length that extend all the way to the surface of the brain (e). Other dendrites surround the cell body (d). Output signals from the neuron travel from the cell body into the axon (a), which splits into several branches (c). The longest branch of the giant pyramidal axon can travel all the way into the spinal cord, a distance that can be more than a meter. Cajal discovered that each pyramidal neuron has thousands of small protrusions, called spines, arising from its dendrites, seen in this drawing as tiny hairs on all of the cell's appendages except for the axon. Each spine acts as an input structure, receiving a contact (synapse) from another neuron (see page 44).

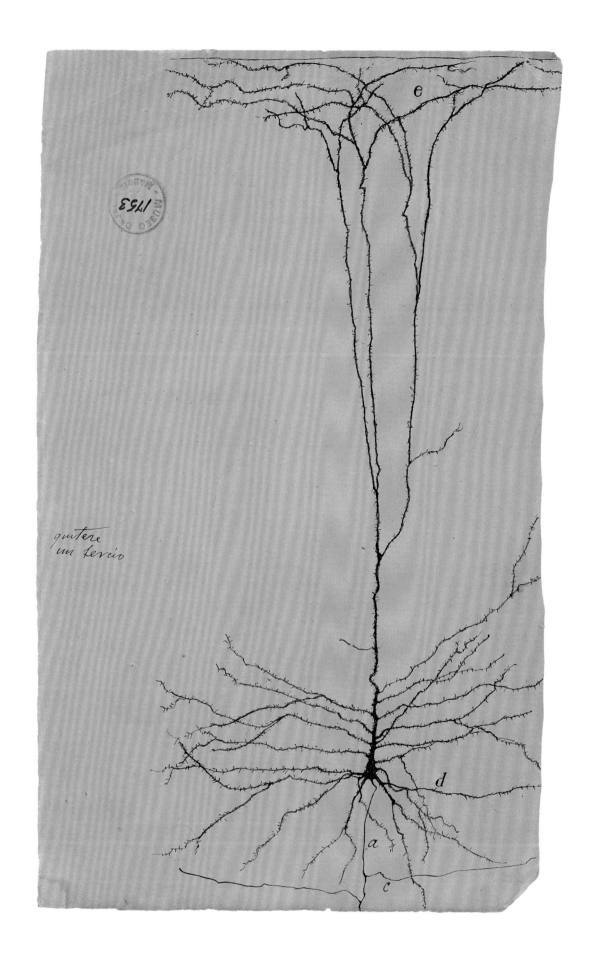

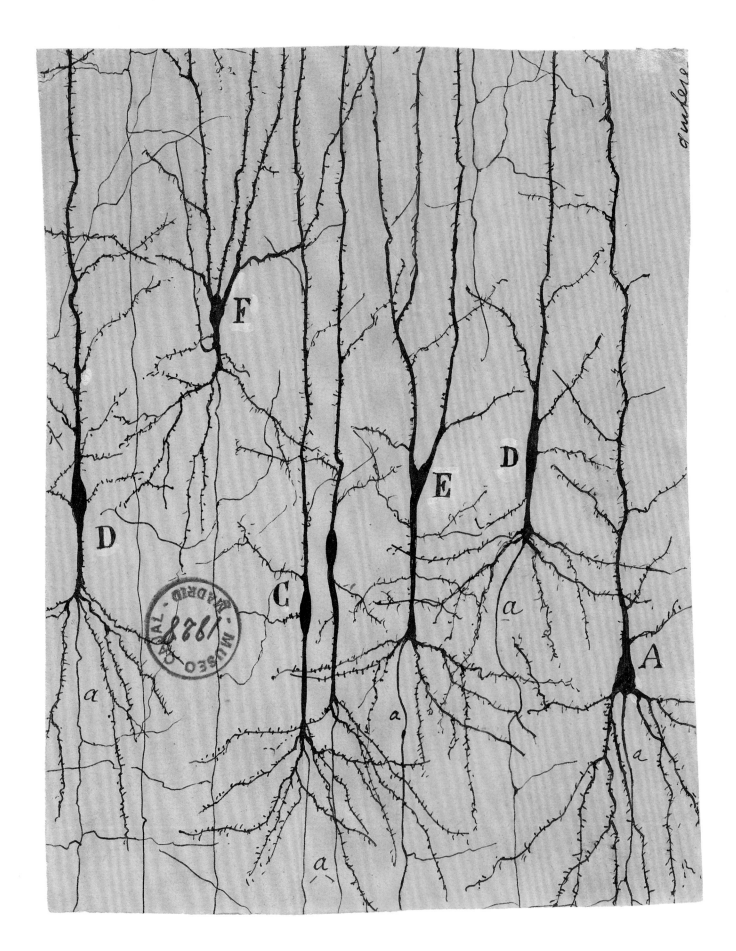

Pyramidal neurons of the cerebral cortex

Cajal stated in 1894, "The cerebral cortex is similar to a garden filled with innumerable trees, the pyramidal cells, which can multiply their branches thanks to intelligent cultivation, send their roots deeper, and produce more exquisite flowers and fruits every day."[1] The beauty of this forest of pyramidal neurons is displayed in this drawing. The profusely branching dendritic trees of the neurons receive and combine information from many other brain areas. Pyramidal neurons in one region of the cortex control the voluntary movements of our bodies, while those in other areas are implicated in cognitive functions and self-awareness. In sensory areas of the cortex, pyramidal neurons process information from the eyes, ears, nose, and skin, and in other areas, the pyramidal neurons combine these different types of sensory inputs.

Pyramidal neurons of the cerebral cortex and their axon pathways

The signals from pyramidal neurons in the cerebral cortex travel to many other parts of the brain and spinal cord. In this drawing of a forest of pyramidal neurons, Cajal shows the pathways followed by the cells' axons, their output appendages. The axons extend downward from the neurons' cell bodies and branch into several daughter axons (called axon collaterals). Some of these axon branches stay within the cerebral cortex, sending information to other neurons in adjacent areas (indicated by arrows in layer A). Other axon branches travel deeper below the brain surface (a, b, c, and d in layer C), sending information to distant parts of the brain.

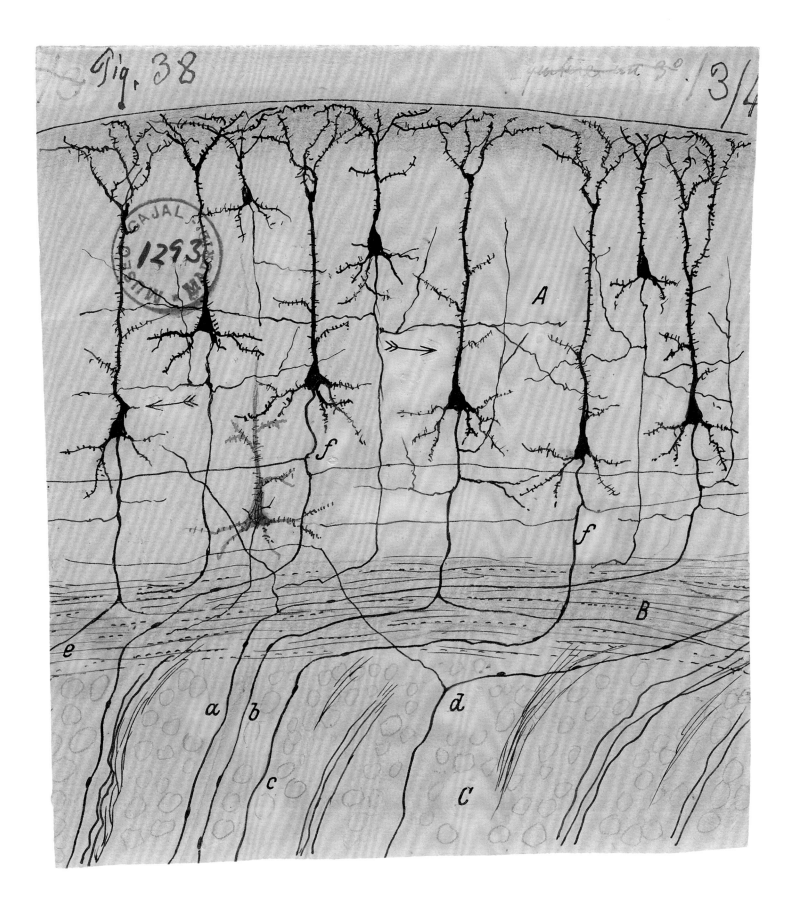

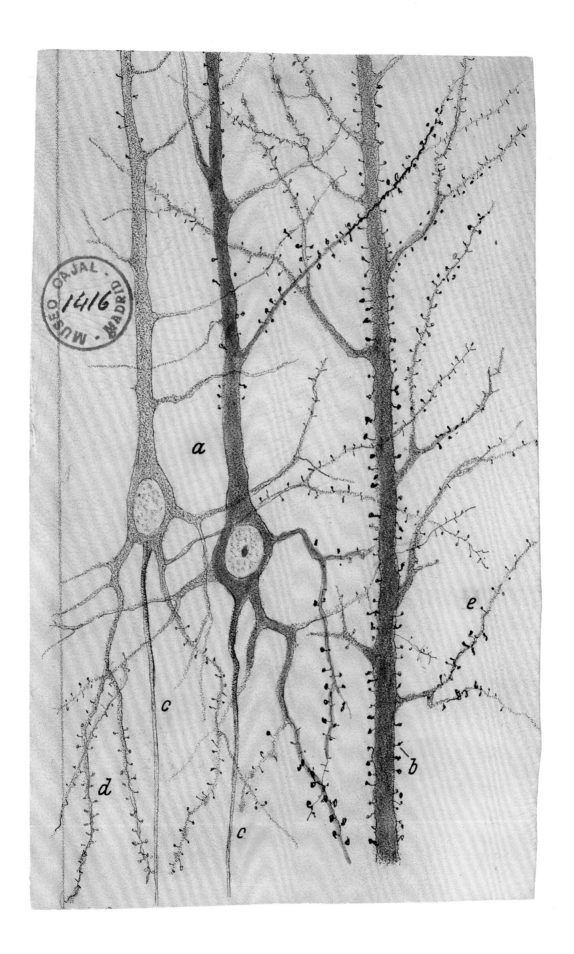

Dendrites of pyramidal neurons of the rabbit cerebral cortex

One of Cajal's important discoveries was the dendritic spine. Cajal observed these spines, the fine hairs that profusely cover the dendrites of many neurons, using the Golgi method to stain the neurons. Many of Cajal's contemporaries believed that these spines were artifacts of the Golgi staining method and did not exist in living cells. To counter this argument, Cajal demonstrated that the spines were also seen when neurons were stained by an entirely different method. This drawing illustrates the multitude of spines on dendrites of pyramidal neurons stained by the Ehrlich methylene blue method, discovered by the famous microbiologist Paul Ehrlich. Results such as this convinced many scientists that dendritic spines were real. Cajal speculated that the dendritic spine was the input structure that received signals from other neurons. We now know that Cajal was correct and that spines receive synaptic inputs that control the electrical responses of neurons. The number and size of the spines present on dendrites can vary with the health of the brain. Spine size and numbers decrease as a result of many diseases associated with cognitive deficits, including Alzheimer's disease, Parkinson's disease, autism, Down syndrome, schizophrenia, and drug addiction.

Pericellular nests around pyramidal neurons in a one-month-old human infant

Cajal discovered a type of neuron in the cerebral cortex whose axons branch profusely and wrap around the cell bodies of pyramidal neurons in the cortex. These branching axons form dense pericellular nests, several of which are shown in this drawing. Although recent research has confirmed Cajal's discovery of these nests, their function remains unknown.

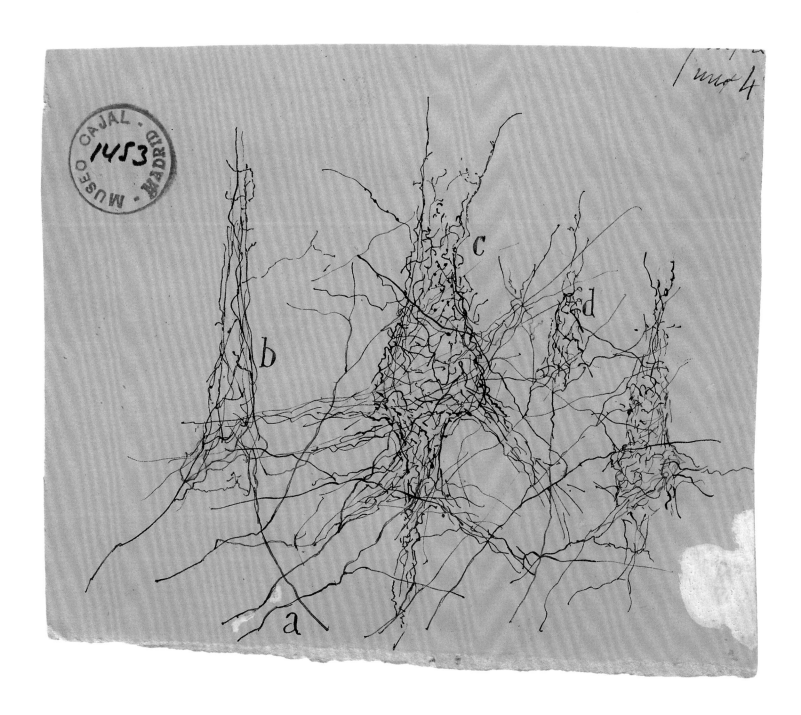

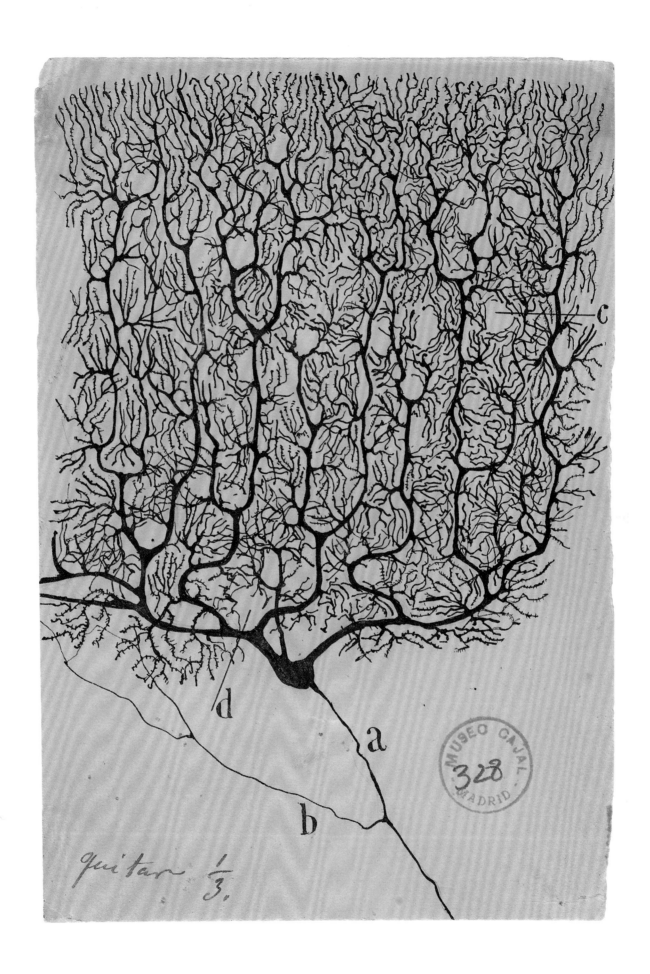

A Purkinje neuron from the human cerebellum

Cajal mused in his autobiography, "In our parks are there any trees more elegant and luxurious than the Purkinje cell from the cerebellum . . . ?"[2] The Purkinje neuron was first observed by Jan Purkinje, the renowned Czech biologist, in 1837. Cajal studied and described these cells in great detail a half century later. The Purkinje neuron is located in the cerebellum, a structure at the back of the brain that lies underneath the cerebral cortex. These cells have an incredibly elaborate dendritic tree structure, making them among the most recognizable neurons in the brain. Unlike pyramidal neurons, Purkinje neurons are not axially symmetric. Their dendritic trees spread out in two dimensions, like a handheld fan. Viewed from the side, they appear flat. Purkinje neurons of the human brain have a more elaborate and complex dendritic tree than those of lower animals.

Purkinje neurons from the pigeon cerebellum

Each Purkinje neuron, two of which are illustrated in this drawing, receives tens of thousands of synaptic inputs onto its elaborate dendritic tree. These inputs come indirectly from our body's sense organs, conveying information about the visual world, head and body position, and the activity of our muscles. By combining all of this information in Purkinje neurons, the cerebellum helps to maintain our bodies in an upright posture when we are standing and is involved in the fine control of our movements.

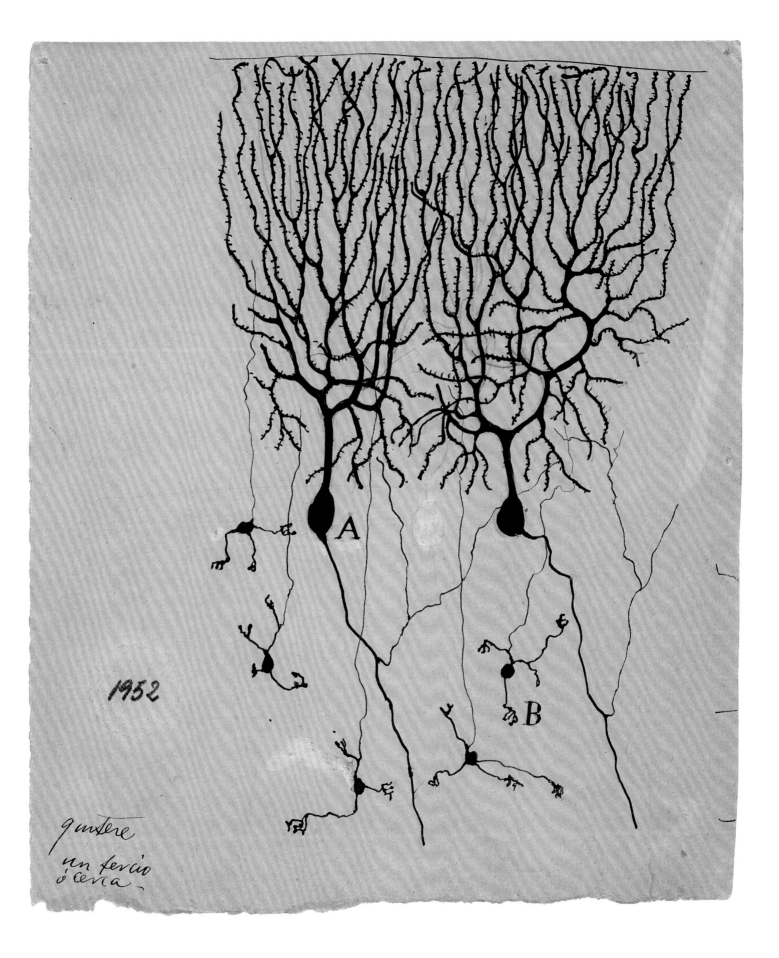

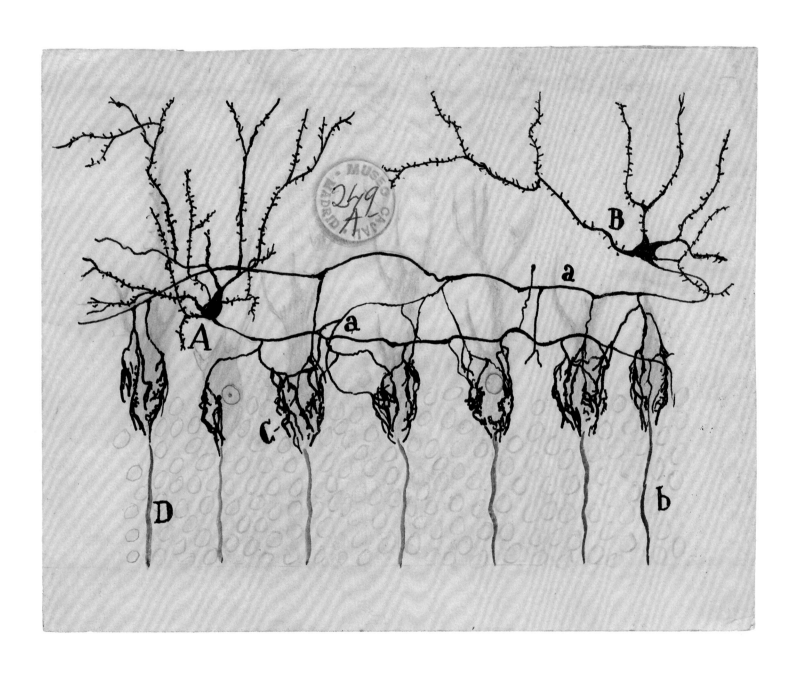

Synaptic contacts in the cerebellum

Cajal used this drawing of neurons in the cerebellum to support his Neuron Doctrine, the theory that the brain is composed of distinct cells rather than a large network of continuously connected cells. The drawing shows a row of seven Purkinje neuron cell bodies and two stellate neurons (A and B), a smaller type of neuron in the cerebellum. Axon branches of the stellate neurons (a) encircle the cell bodies of the Purkinje neurons, making multiple synaptic contacts. Cajal noted that axons of the stellate neurons made distinct contacts (synapses) onto Purkinje neurons, but did not fuse with them, supporting the Neuron Doctrine of discrete, separate neurons in the brain. The contacts of these stellate neurons form basketlike networks of appendages surrounding Purkinje neuron cell bodies, which is why these cells are known as basket neurons.

Stellate neurons in the cerebellum

Stellate neurons are one of several types of interneurons in the cerebellum, neurons that signal to other neurons within their immediate neighborhood but do not send signals to distant brain regions. Some stellate neurons in the cerebellum contact and form synapses on Purkinje neurons (see page 52). These synapses are inhibitory, making the Purkinje neurons less likely to generate electrical responses.

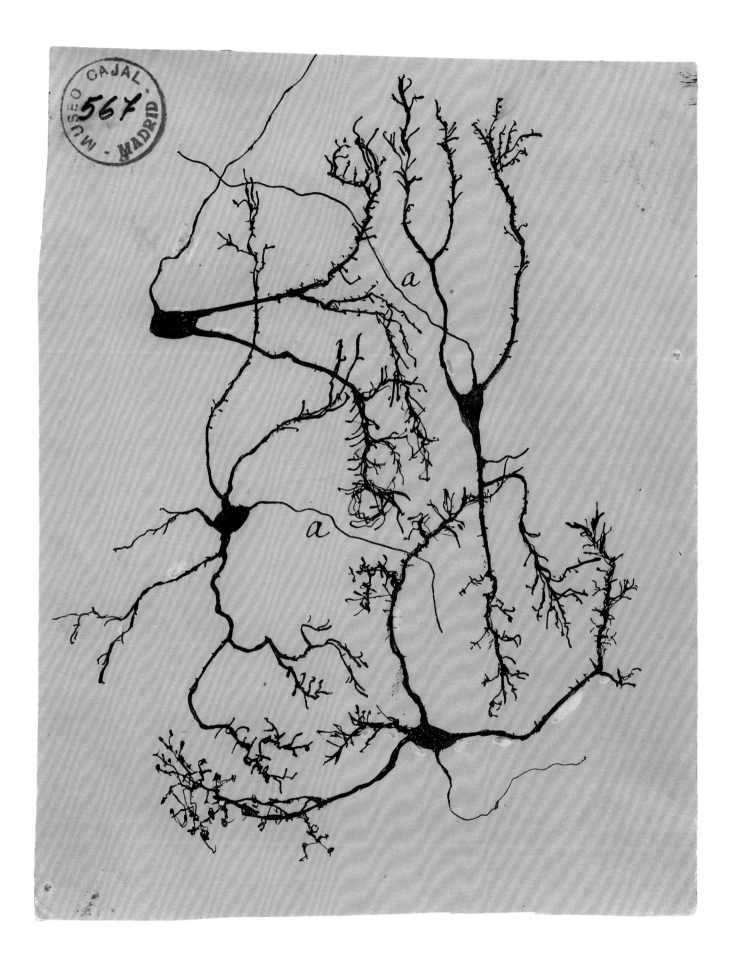

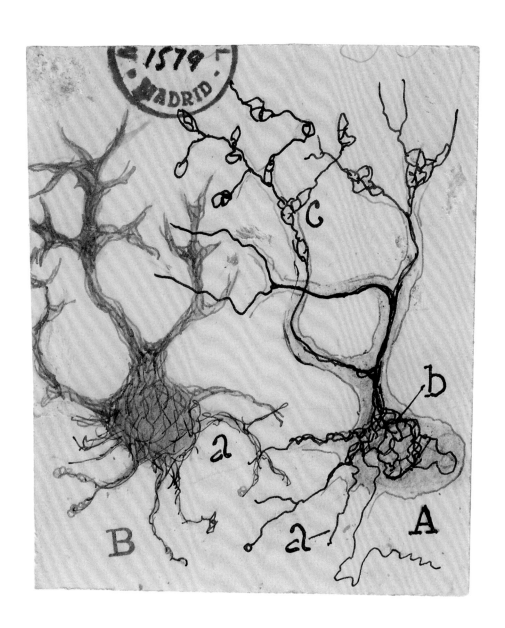

The internal structure of neurons

Neurons and other cells in our bodies have an internal skeleton to maintain their shape. A cell's skeleton (called the cytoskeleton) is made of proteins that form long fibers and tubes (neurofilaments and microtubules). Although Cajal did not know the chemical composition of these skeletal components, he invented a staining technique (the reduced silver-nitrate method) to reveal their shapes within neurons. In this drawing, the cytoskeleton is shown in the right-hand neuron by dark lines within the cell body and the dendrites. The drawing illustrates a technique that Cajal often employed. He unified, in a single drawing, observations made at different times or obtained using different methods, thus allowing him to illustrate a concept or a hypothesis. In this drawing, the neuron on the left is from a brain stained with one method, while the neuron on the right comes from a different brain stained with a second method.

Neurons in the frog's small intestine

Our nervous system is made up of many parts, including the brain, the spinal cord, and the autonomic nervous system, which controls involuntary (automatic) functions of our bodies. Our gut has its own nervous system, the enteric nervous system, that controls the muscles of the esophagus, stomach, and intestines. The enteric nervous system is not a minor affair. It contains one hundred million neurons, at least as many as the entire spinal cord! This drawing shows the neurons of the myenteric plexus, a part of the enteric nervous system that lies between muscle layers in the gut.

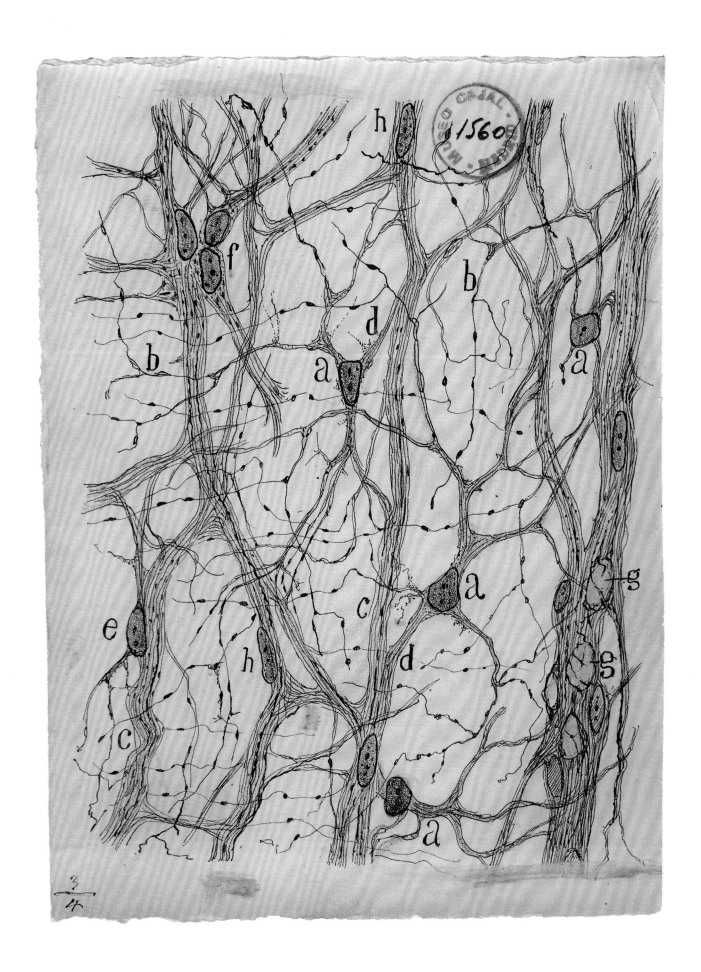

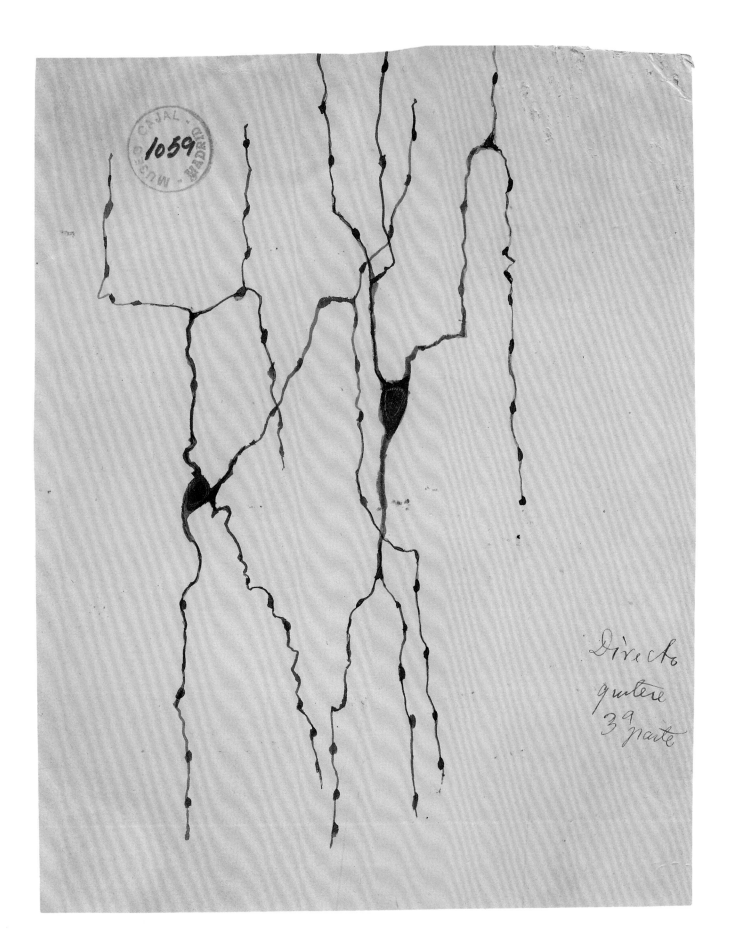

Cajal's neurons in the gut

The enteric nervous system (see page 59) is made up of many types of neurons. Cajal discovered one type of enteric neuron, which has been named in his honor—"the interstitial cells of Cajal." Recent research has shown that these neurons play a crucial role in the functioning of our digestive system. They generate repetitive electrical signals that drive peristalsis, the slow cyclic motion that moves food through our intestines.

Neurons in the superior cervical ganglion

The superior cervical ganglion is part of the autonomic nervous system that controls our immediate response to danger. The ganglion, a structure that contains a large group of neurons and lies outside of the brain proper, controls the responses of our head and neck, including blood flow in our brain and the production of tears and saliva. In the drawing on the right, several superior cervical ganglion neurons and their complex dendrites are shown. In the drawing on the left, a large superior cervical ganglion neuron is surrounded by axons that wrap around the neuron's cell body and dendrites. Cajal believed that these axonal wrappings, which he called nests, were due to old age or disease. Modern research has confirmed Cajal's conclusions.

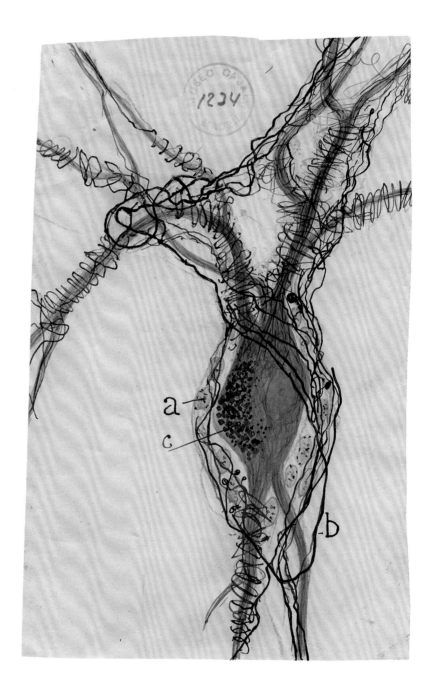

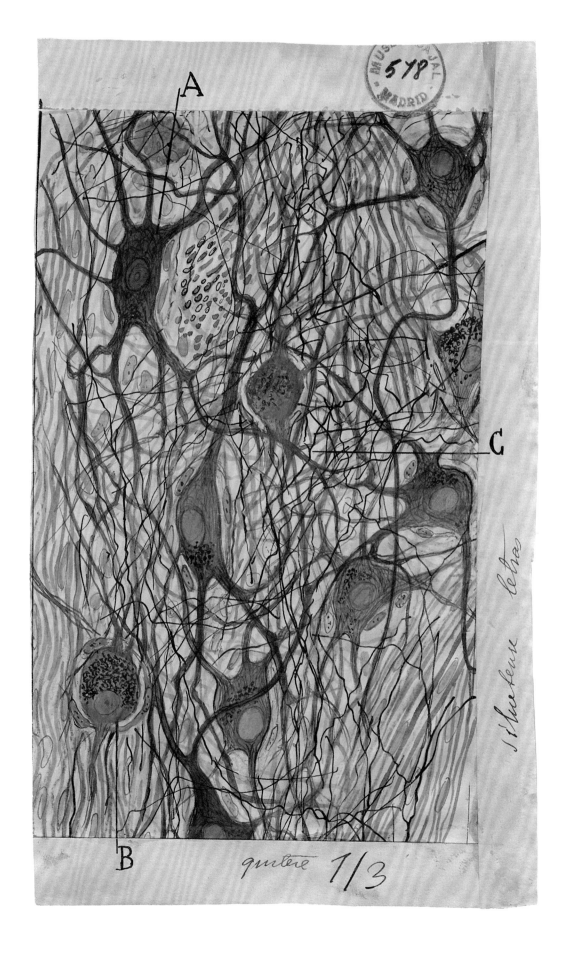

Glial cells of the cerebral cortex of a child

Astrocytes, the most common type of glial cell in the brain, come in a variety of shapes. Cajal illustrates some of this variety in this drawing of astrocytes near the surface of the cerebral cortex. Classical star-shaped astrocytes, from which this class of cells gets its name, are seen in the deeper layers of the cortex (cells E, F, G, H, I, J, K, and R). Some of these astrocytes (cells G, I, and J) make contacts with a blood vessel (V). Other, more elongated astrocytes near the surface of the brain (cells A, B, C, and D) make contacts with the brain surface, forming the brain lining known as the glia limitans.

Astrocytes in the hippocampus of the human brain

Cajal beautifully summarizes many of the properties of astrocytes in this drawing of the hippocampus of a man three hours after death. Holding center stage in the middle of the drawing is an astrocyte with its classical star-shaped appearance. Some of its appendages contact a neuron (the large, lightly shaded cell on its right). Other appendages contact a blood vessel (F) to its left. A second astrocyte (A) hugs a neuron, emphasizing the intimate relation between astrocytes and neurons in the brain. A third astrocyte (B) is caught in the act of dividing into two daughter cells. This is important because astrocytes, but few neurons, are able to divide in the adult brain. A fourth astrocyte (E) shows signs of degeneration.

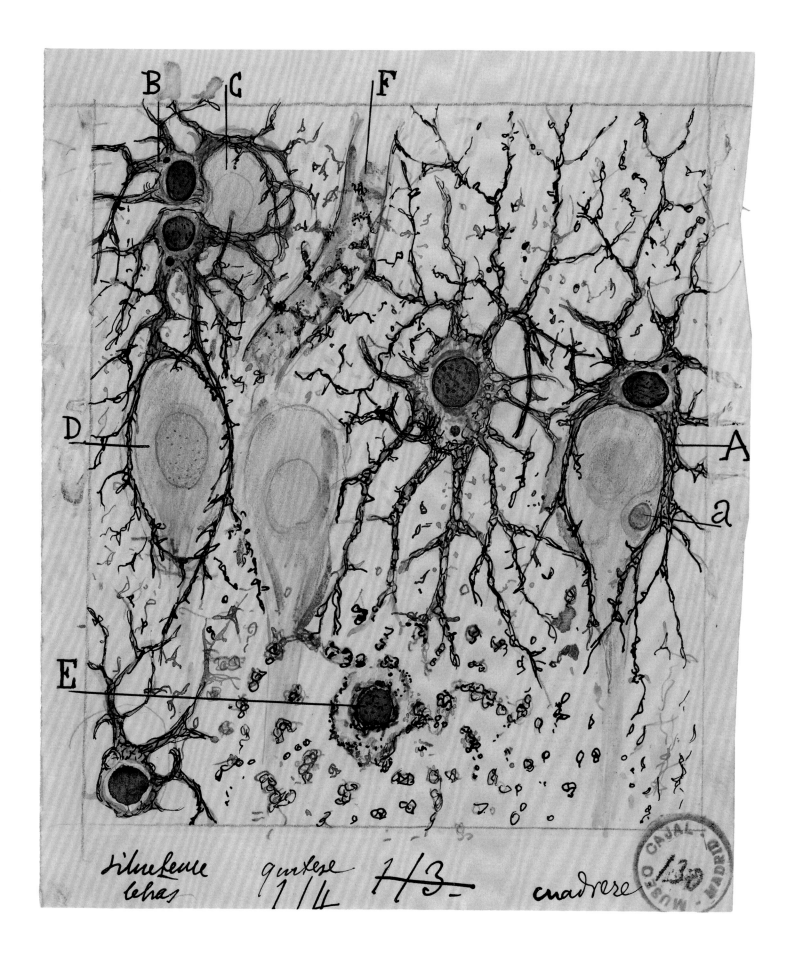

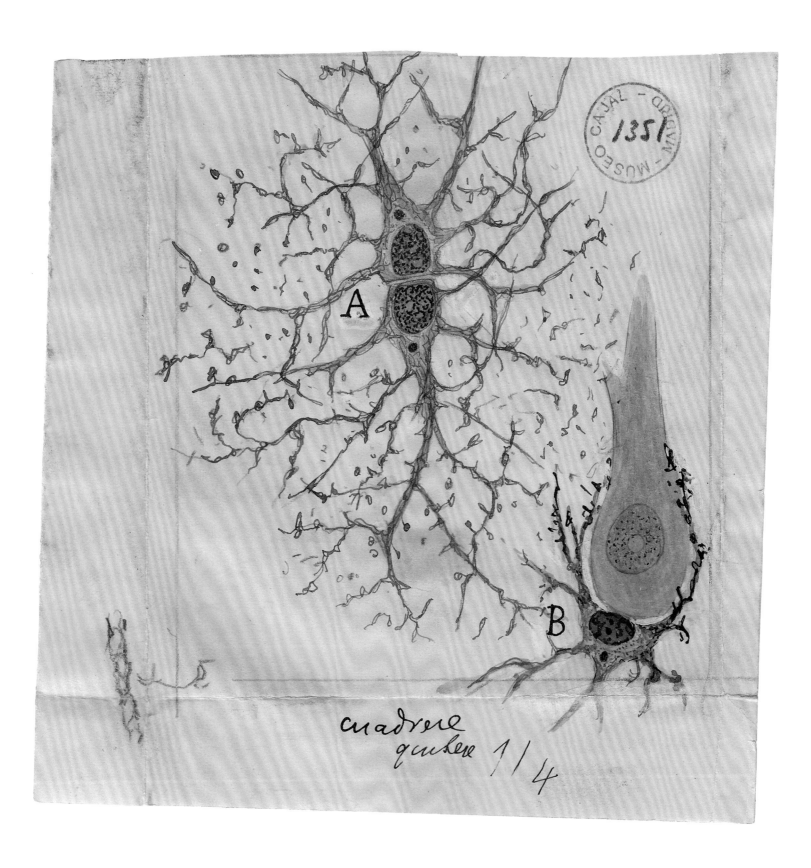

Twin astrocytes in the cerebral cortex of the dog

As in the illustration on the previous spead, Cajal emphasizes important astrocyte properties in this drawing. One astrocyte (A) is dividing into two daughter cells. Cajal called these cells twin astrocytes. Astrocytes are able to divide throughout life and particularly following an injury to the brain (see page 175). Neurons in the adult brain, in contrast, rarely divide. The appendages of a second astrocyte (B) envelop a neuron. This close association between astrocytes and neurons led Cajal to suggest that astrocytes might regulate the electrical responses of neurons. Within the past twenty years, research has proved Cajal to be correct; astrocytes can modulate electrical signaling in neurons. This and the previous drawing illustrate an important technique that Cajal used in his drawings. Although he knew that astrocytes were substantially smaller than neurons, he drew the astrocytes oversize to draw attention to their leading role in the story he wanted to tell. Their prominence is further emphasized by his coloring the astrocytes darkly and the neurons lightly.

Glial cells surrounding pyramidal neurons in the human hippocampus

According to Cajal, "the human cortex differs from that of other animals not only in the huge amount of glandular cells [astrocytes] that it contains, but in their smallness and the wealth of the interstitial glial plexus [astrocyte appendages]."[3] This observation has been confirmed in recent studies, which show that astrocytes in the human brain are more numerous and more complex, having far more branched appendages, than astrocytes of simpler animals. Some scientists have speculated that the number and complexity of astrocytes in the human brain is partially responsible for our superior intelligence.

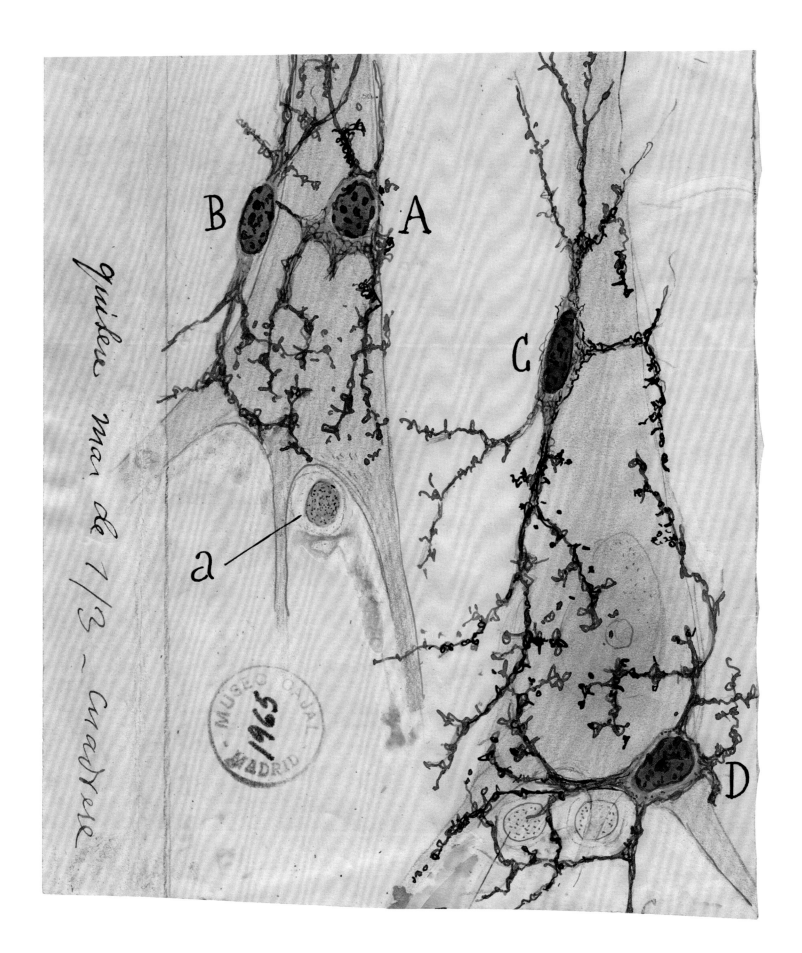

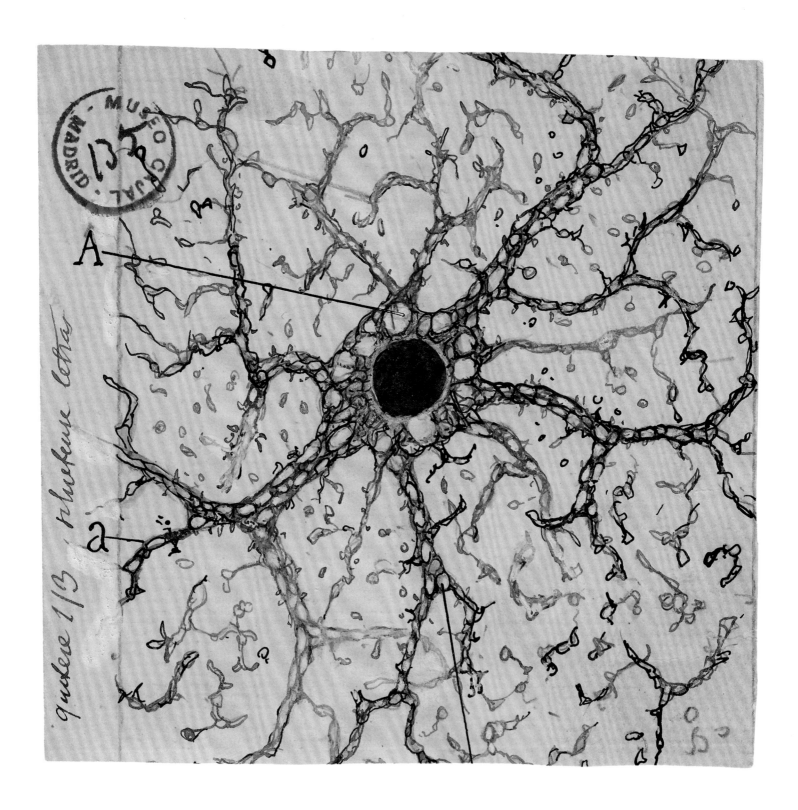

An astrocyte in the human hippocampus

Cajal speculated that the astrocytes associated with neurons in the cerebral cortex "constitute a vast endocrine gland," releasing chemical signals (hormones) "associated with brain activity."[4] Here again, Cajal was close to the mark. Although they don't release hormones per se, astrocytes do release chemicals (gliotransmitters) onto synapses that modulate the electrical responses of neurons. Astrocytes also release other chemicals onto blood vessels that regulate blood flow in the brain. Illustrated in this drawing is an astrocyte in the hippocampus with its multiple branched appendages.

Gray matter astrocytes in the spinal cord

Cajal proposed that the endfoot appendages of astrocytes pull on the walls of blood vessels when the brain is active, causing them to dilate and to increase blood flow: "Every astrocyte of the white or gray matter is provided with a sucking apparatus or perivascular pedicle [endfoot]. . . . The purpose of these elements is to provoke, by contraction of such appendages, local dilation of the vessels, and thus increased blood flow linked to the intensity of the mental processes."[5] A century later, research has shown Cajal to be largely correct. Although astrocyte endfeet do not contract, they do release chemicals that cause blood vessels to dilate, bringing extra blood and nutrients to active neurons. This drawing illustrates the contacts between astrocytes (B) and neurons (a and b) and between astrocyte endfeet (c) and a blood vessel (V).

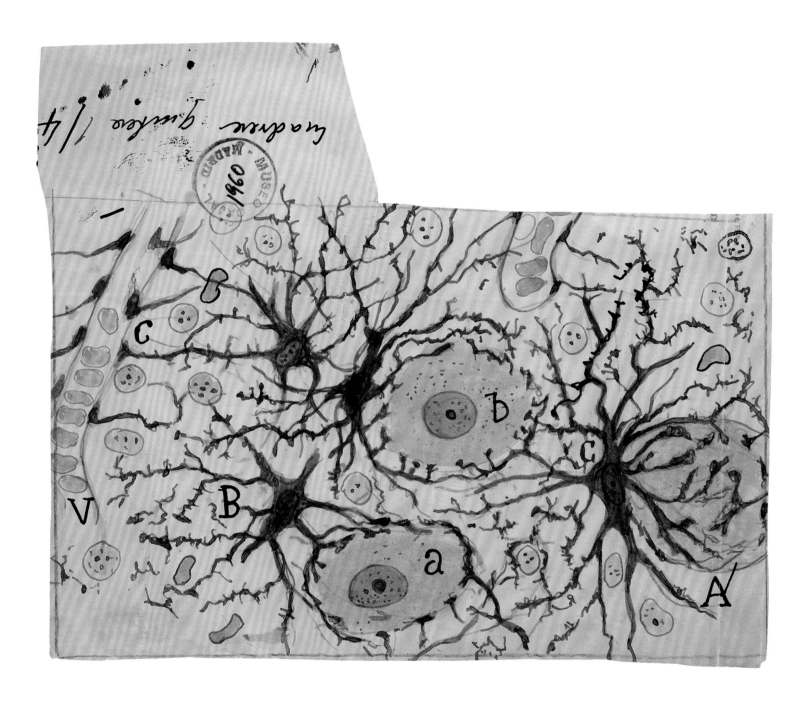

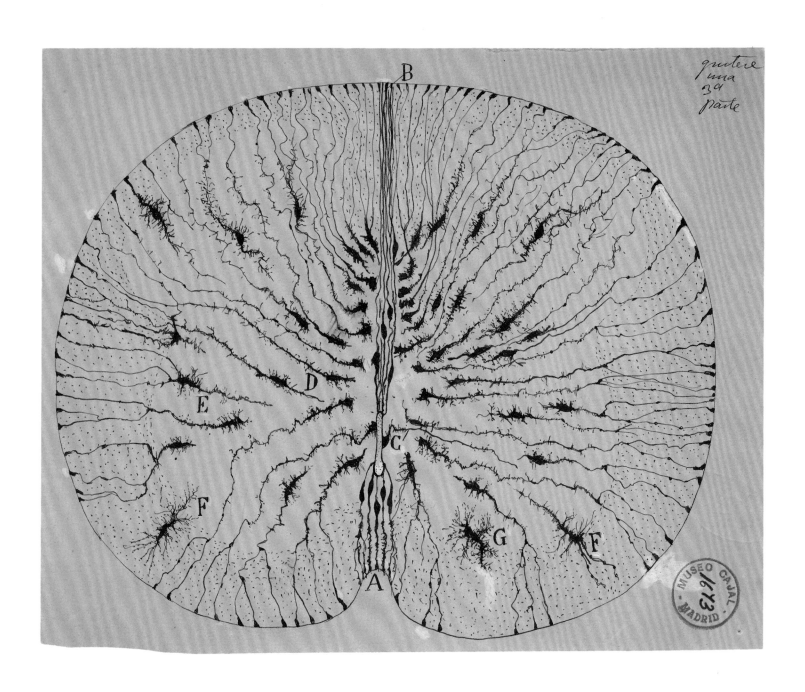

Glial cells of the mouse spinal cord

Cajal studied the development of the brain as well as the structure of the adult brain. Illustrated in this drawing are the glial cells of the developing spinal cord. Most of the glial cells pictured are super-elongated cells, stretching all the way from the center of the spinal cord to its surface (cell D, for instance). These cells are now known as radial glial cells. Cajal correctly proposed that these cells transform into astrocytes (cells F and G) as the animal matures. Similar radial glial cells are present in the developing brain as well as the spinal cord. In a completely unexpected finding, recent research has shown that most neurons in the brain, as well as astrocytes, are derived from these radial glial cells.

Cells in the optic lobe of the fly

The nervous systems of insects and humans have many similarities. Both contain neurons possessing dendrites and axons. Both insect and human neurons generate electrical impulses that transmit information between distant regions of their respective brains. The insect and human nervous systems also possess glial cells that serve comparable functions in their respective brains, including glial support of neuronal nourishment and the modulation of the neurons' electrical responses. This drawing illustrates part of the optic lobe of the fly, the part of the insect's brain that processes visual information from its retina. This structure has its own glial cells, and Cajal highlights these glial cells (A) in this illustration.

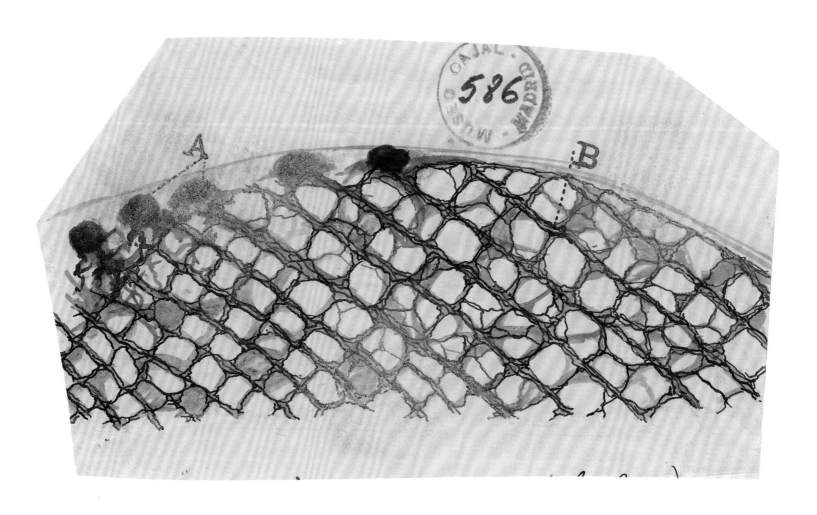

reducido á la
en una 3ª parte   2/3

Muscle cells from the leg of the scarab beetle

In addition to studying many species of vertebrate animals, including human, dog, chicken, lizard, frog, and fish, Cajal also studied invertebrates. This drawing illustrates the structure of the leg muscle of a beetle. As shown in the lower two sketches, insect muscle cells have a repeating, banded structure. Although not known in Cajal's time, two proteins (actin and myosin), present in each repeating unit in the muscle cell, slide past each other, causing the muscle to shorten. The same proteins and a similar muscle structure are responsible for muscle contraction in humans.

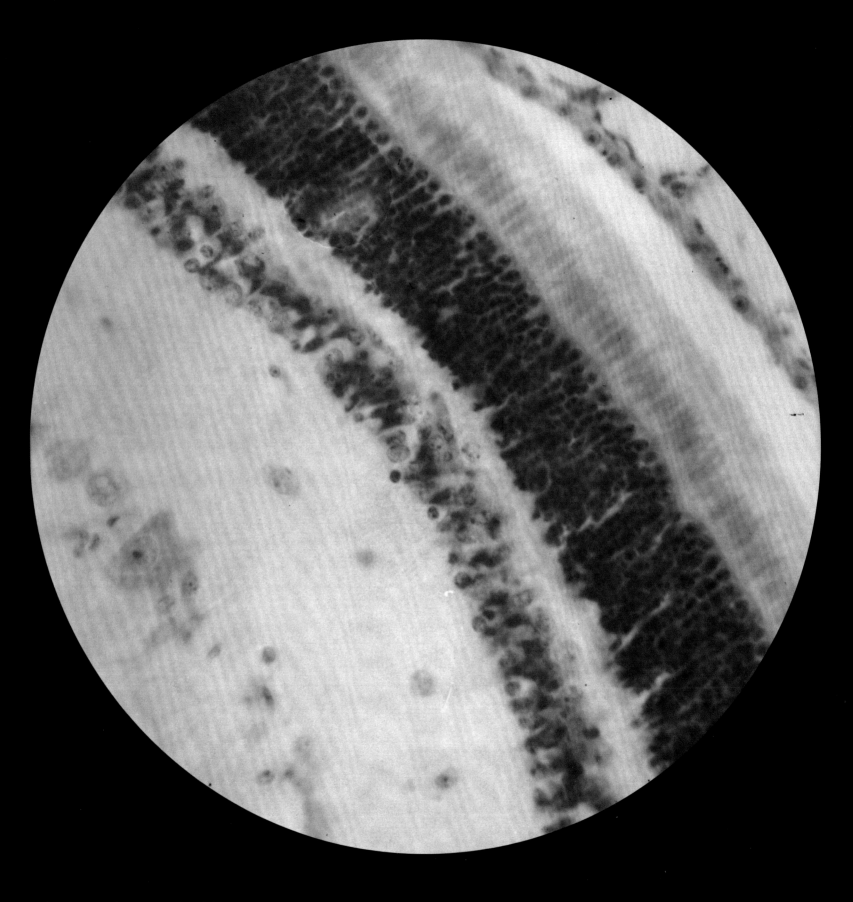

Retina of the ox, microphotograph by Cajal

SENSORY **SYSTEMS**

The brain itself cannot perceive the external world. For that, it relies on our specialized sense organs, including the eye, the inner ear, and the nose. The retina, a paper-thin structure at the back of the eye, is the light-sensitive tissue responsible for vision. The retina is an integral part of the central nervous system and is composed of the same types of cells as the brain—neurons and glial cells. Images from the outside world are focused by the eye's cornea and lens onto the retina, which converts these images into electrical signals that travel to the visual centers of the brain. The brain processes these signals into the images that we see. These basic principles were known to Cajal and his colleagues over a century ago.

Cajal was fascinated by the retina and studied its structure in many species, including humans, other mammals, birds, fish, and insects. He reminisces in his autobiography, "[T]he retina [is] the oldest and most persistent of my laboratory loves. . . . [L]ife never succeeded in constructing a machine so subtly devised and so perfectly adapted to an end as the visual apparatus. . . . I felt more profoundly than in any other subject of study the shuddering sensation of the unfathomable mystery of life."[6] Many of Cajal's most important conceptual breakthroughs were aided by his studies of the retina.

One of Cajal's key contributions to neuroscience was his deduction of the direction of information flow within the brain. Individual neurons are composed of a cell body, a series of thick branching appendages called dendrites, and a single long, thin appendage called the axon. Cajal deduced that information flows from the dendrites to the cell body and then to the axon.

Cajal states in his autobiography, "[W]e observe that in the [retina] . . . the thick processes of the cells [dendrites] are always directed towards the external world and evidently conduct towards the cell body, while the axon . . . is directed towards the central nervous [system]. Proceeding by induction, it was natural to attribute similar dynamic properties to the dendrites of the multipolar neurons in the cerebral hemisphere, the cerebellum, and the spinal cord."[7]

In other words, Cajal's understanding of the retina suggested to him that information from the outside world is received by the dendrites of the retina's neurons.

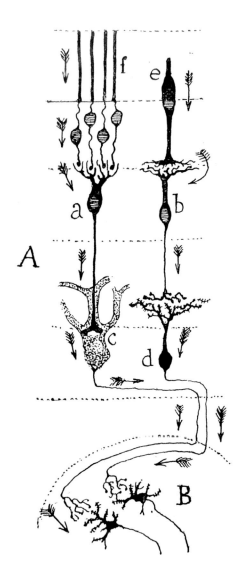

Electrical signals then travel to the cell body of the neuron and finally out through the axon, which makes contacts with other neurons. Cajal then generalized this concept of information flow to neurons throughout the brain.

Cajal's observations on the flow of information in retinal neurons are beautifully summarized in his drawing shown at the left. Arrows in the drawing indicate that information flows from the light-sensitive cells (photo-receptors) at the top of the retina (e and f) to the dendrites and cell bodies of intermediate cells (a and b) and then to the dendrites, cell bodies (c and d), and axons of cells at the bottom of the retina. Information then travels down the axons of these cells to the brain (B).

It is remarkable that Cajal could infer the direction of information flow within and between neurons simply by studying the structure of cells in the retina, with additional observations in the brain and spinal cord. Without actually recording the electrical signals within neurons, which was not technically feasible in Cajal's time, he was able to deduce how information travels throughout the nervous system.

SENSORY SYSTEMS

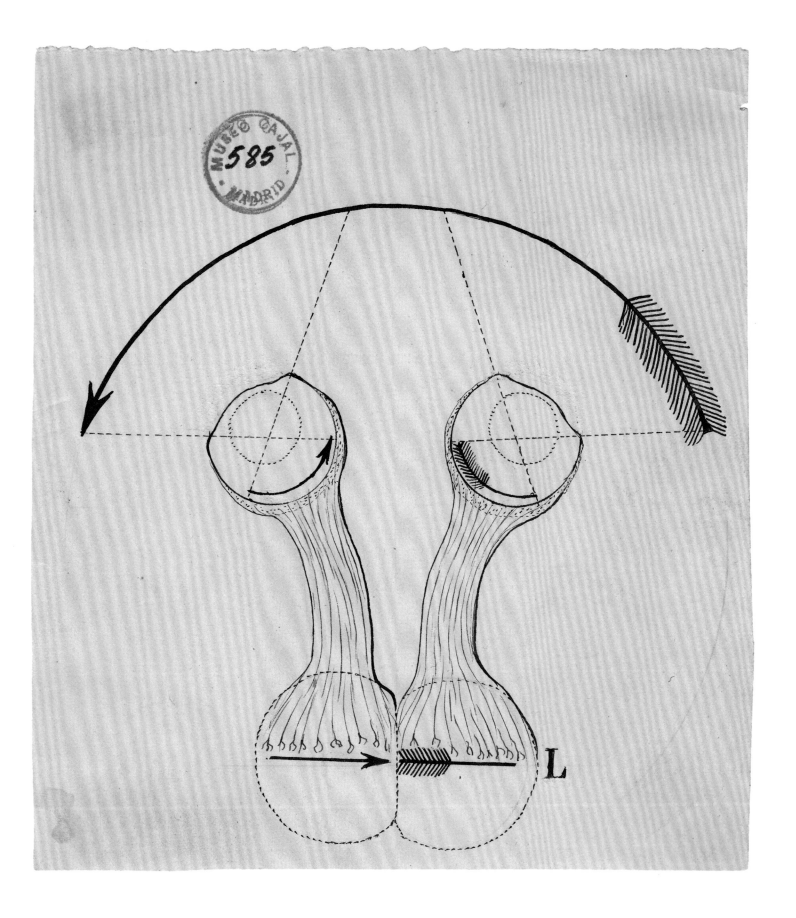

Diagrams indicating how information from the eyes might be transmitted to the brain

It was well known in Cajal's time that much of the information from the right eye traveled to the left side of the brain and vice versa. Cajal puzzled over why this was true in all animals. He imagined our two eyes looking at an arrow, as illustrated in these drawings. He reasoned that if information from the two eyes did not cross as it traveled to the brain, as shown in the drawing opposite, then a unified representation of the visual world could not be created in the brain. That is, the representation of the arrow in the brain (L) would not be continuous. On the other hand, if information from the eyes did cross, as illustrated in the right-hand drawing, a unified representation would result (arrow Rv). Cajal summed up his reasoning in his autobiography, stating, "[I]n order that the mental perception may be unified and may agree exactly with the external reality, or, in other words, in order that the image conveyed through the right eye may be continuous with that conveyed through the left eye, the intercrossing of the optic paths from side to side is quite necessary."[8]

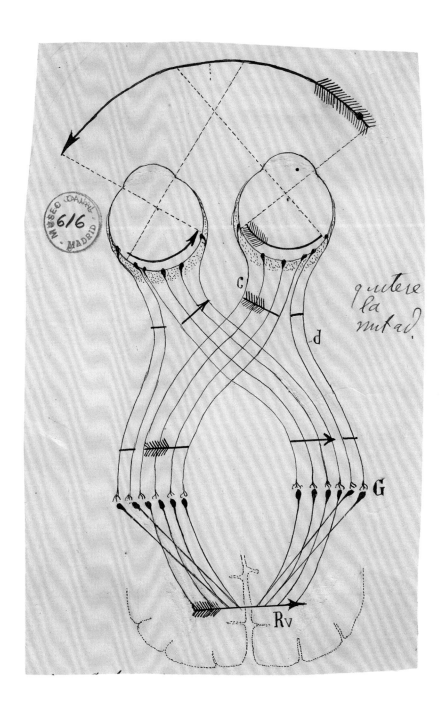

SENSORY SYSTEMS
87

Cells in the retina of the eye

Cajal summarizes in this drawing all of the important classes of cells in the retina, the light-sensitive tissue at the back of the eye. He emphasizes as well the many structural layers of the retina, which are labeled with capital letters at the right. In Cajal's time, it was known that light falling on the retina activated light-sensitive neurons (photoreceptors), located in layers B, C, and D. These cells convert light into electrical signals. As Cajal inferred, these signals are transmitted by contacts (synapses) to relay neurons in layer F (bipolar cells) and then to a third type of neuron in layer H (ganglion cells), which transmits the visual information from the retina to the brain. Other neurons in layer F (horizontal cells and amacrine cells) contribute to the first steps in the processing of visual information even before the signals reach the brain. The cells at the right, labeled ñ and o, are glial cells, non-neuronal cells that assist neurons in the processing of visual signals. By placing them to the side, Cajal draws a clear distinction between the glial cells and the neurons. Cajal has turned one of the glial cells, the astrocyte (o), on its side so that we can see its shape better. As in a Cubist painting, he is showing us multiple views in a single image.

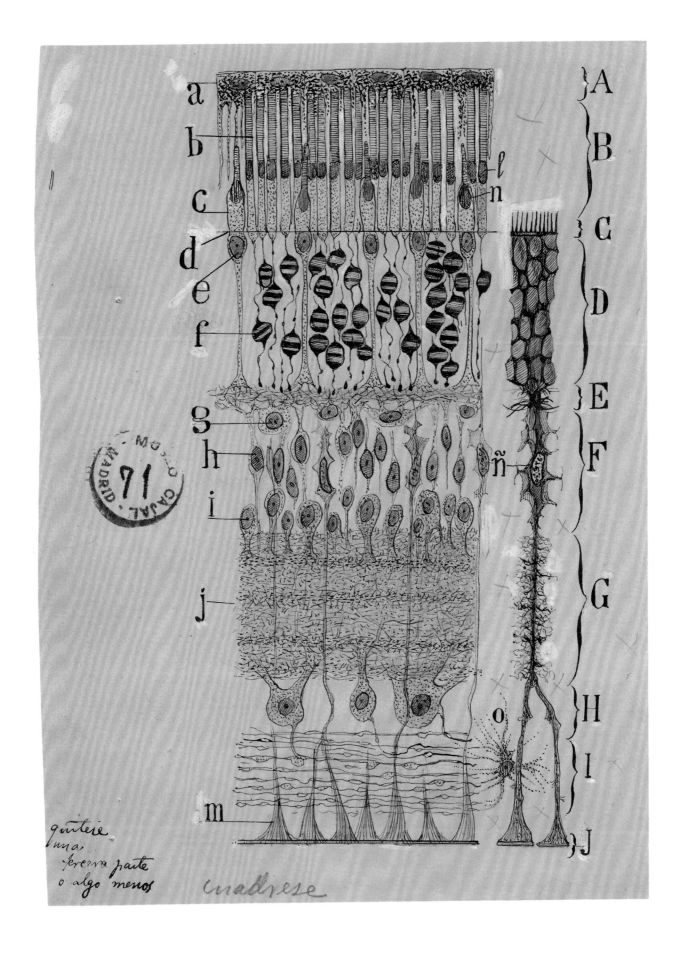

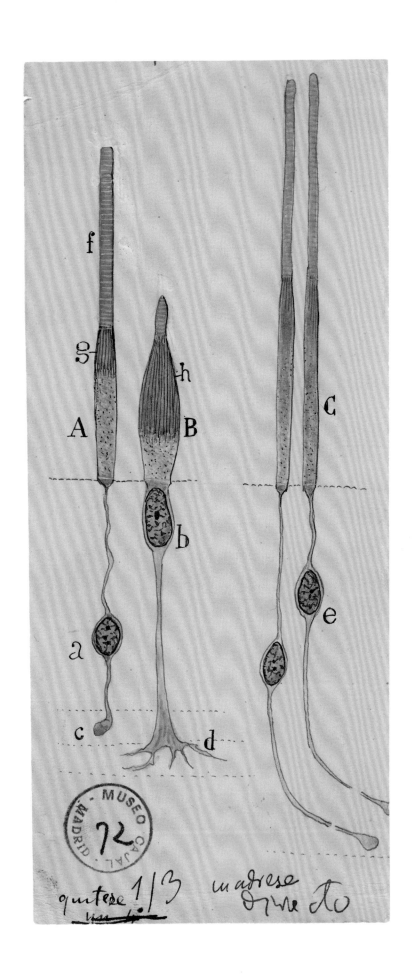

Rods and cones in the human retina

Cajal distinguishes between the two types of light-sensitive photoreceptor neurons in the retina, which convert light into electrical signals. Cell A is a rod photoreceptor, while cells B and C are different subtypes of cone photoreceptors. Cajal knew that the rods are responsible for seeing in dim light, where colors cannot be distinguished, and that cones are responsible for color vision in bright light. The lowercase letters indicate different parts of the photoreceptor cells, including their light-sensitive appendage (f, called the outer segment), their cell bodies (a, b, and e), and their contacts (synapses) with other neurons (c and d).

Cone photoreceptor cells of the pigeon retina

Cajal was fascinated by the variety and specializations found in the retinas of different types of animals. In this drawing he illustrates the light-sensitive photoreceptors of the pigeon retina. Pigeons are active during the day and their retinas have lots of cone photoreceptors, which are sensitive to bright light. Four cone photoreceptors (A) are shown in this drawing. Cajal highlights a specialized feature of the cones of birds, a colored oil droplet (a), the small circular structure just beneath the light-sensitive outer segment appendage of the cone (b, c, and d). As Cajal knew, cone oil droplets come in six different colors that filter the light before it hits the light-sensitive outer segments, giving birds a greater range of color vision than we have. Cajal also illustrates a retinal pigment epithelium cell (B) in the drawing. These cells contain black pigment and form a continuous sheet behind the retina, shielding the retina from stray light.

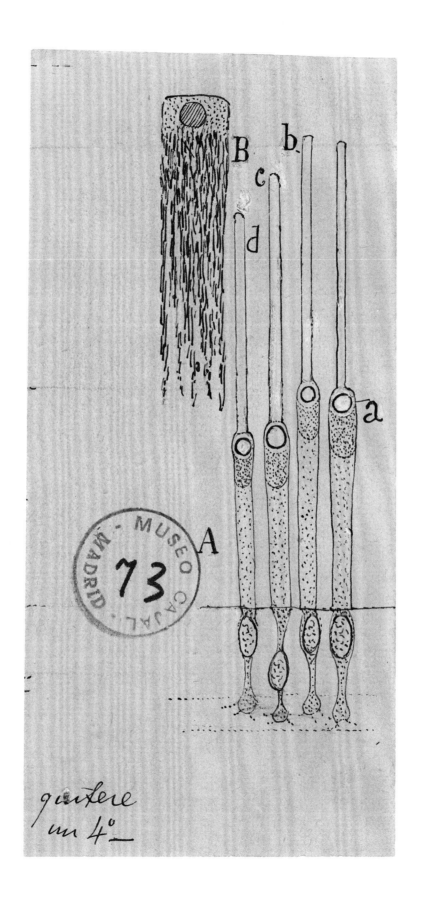

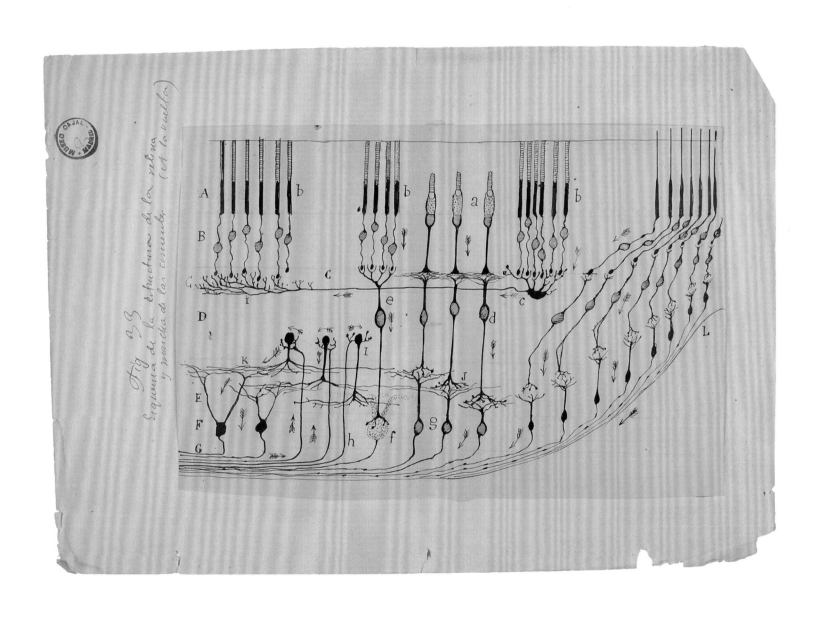

Structure of the retina indicating the flow of information

Cajal deduced the direction of information flow in neurons by examining their structure and knowing their inputs and outputs. He illustrates the pathways of information flow in the retina in this drawing, which not only provides accurate structural details, but includes arrows to show the order in which visual information is processed. Light first stimulates the light-sensitive photoreceptor neurons (layers A and B). As Cajal knew, information is then transmitted to relay neurons (layer D) and then to the output neurons of the retina (layer F). Electrical signals traveling through the long, thin axons of these output neurons (layer G) transmit visual information to the brain (not shown). The retina is thinnest in its central portion (the fovea), shown on the right of this drawing (L). As Cajal observed, the light-sensitive photoreceptors of this region, by being densely packed, give us our sharpest vision.

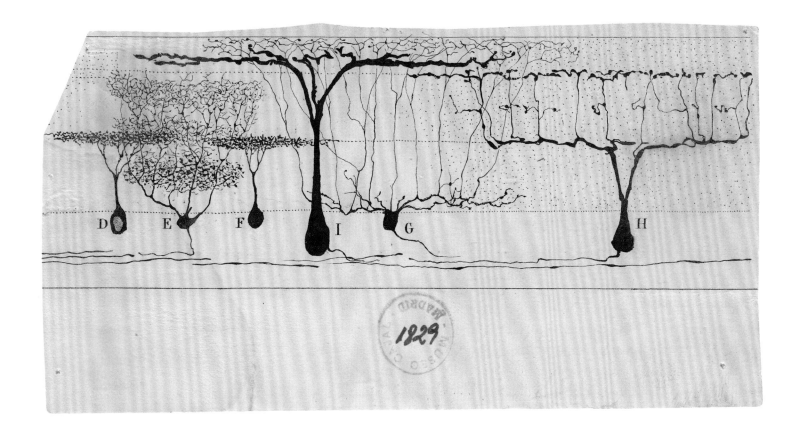

Retina of the lizard

Cajal recognized multiple shapes and sizes among the output neurons of the retina, the ganglion cells, which transmit electrical signals from the retina to the brain. These output neurons differ in the size and shape of their dendritic trees, which receive input signals from other retinal neurons. Cajal correctly grouped these cells together as a single neuronal class because they all have axons (the long, thin appendages at the bottom of the drawing) that travel from the retina to the brain. We now know that the different shapes of these output neurons indicate their specialized functions. Some of these neurons signal the brain when an object in the outside world is moving, others when the light gets brighter, and still others when a red or a green object is present.

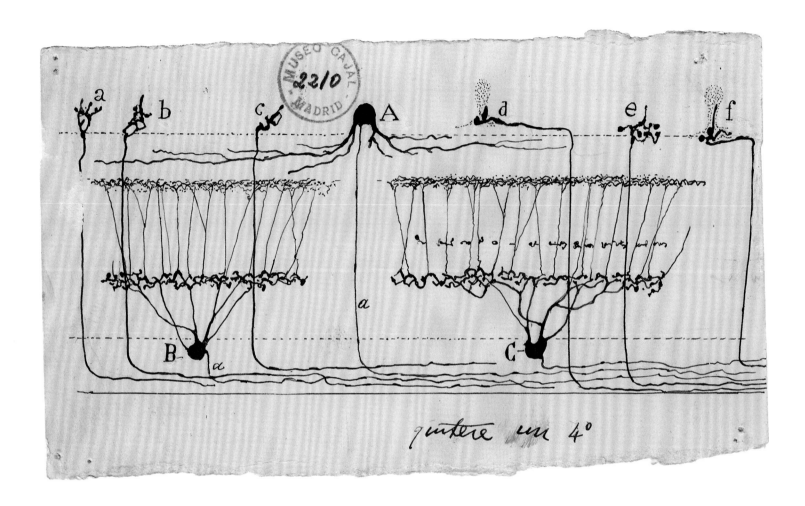

Retina of the sparrow

As Cajal showed in his many beautiful retinal drawings, the output neurons of the retina, the ganglion cells, come in many sizes and shapes. One such subtype of ganglion cell, which Cajal called the bistratified ganglion cell, is illustrated in this drawing (cells B and C). Its dendritic tree is thickly branched in two layers of the retina. In addition, Cajal recognized and recorded in this drawing a rare type of retinal neuron. The cell body of this neuron is in the brain (not shown in the drawing) and its axon travels to the eye to make contacts in the retina. These neurons, labeled a, b, c, d, e, and f, send information from the brain back to the retina and are found in mammals as well as birds. Even today, a century after Cajal described them, we do not know the precise function of these neurons.

SENSORY SYSTEMS

Superior colliculus of the kitten

Visual signals from the eye travel to the cerebral cortex, where the signals are processed and are perceived as objects. Visual signals also travel from the eye directly to the midbrain, where they end in a structure called the superior colliculus, pictured in this drawing. Cajal was the first to describe the branched endings of the axons that travel from the eye to the superior colliculus. This drawing shows bundles of these axons coming from the eye (A) and their endings in different layers of the colliculus (a, b, and c). Based solely on the anatomical pathways, Cajal concluded that this midbrain structure mediated a visual reflex that controlled eye movements. Modern research has confirmed Cajal's speculation.

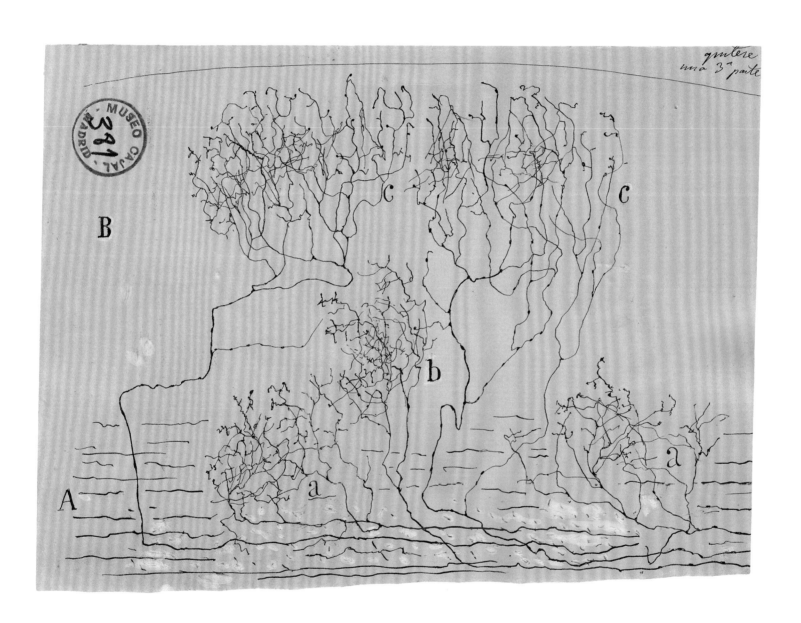

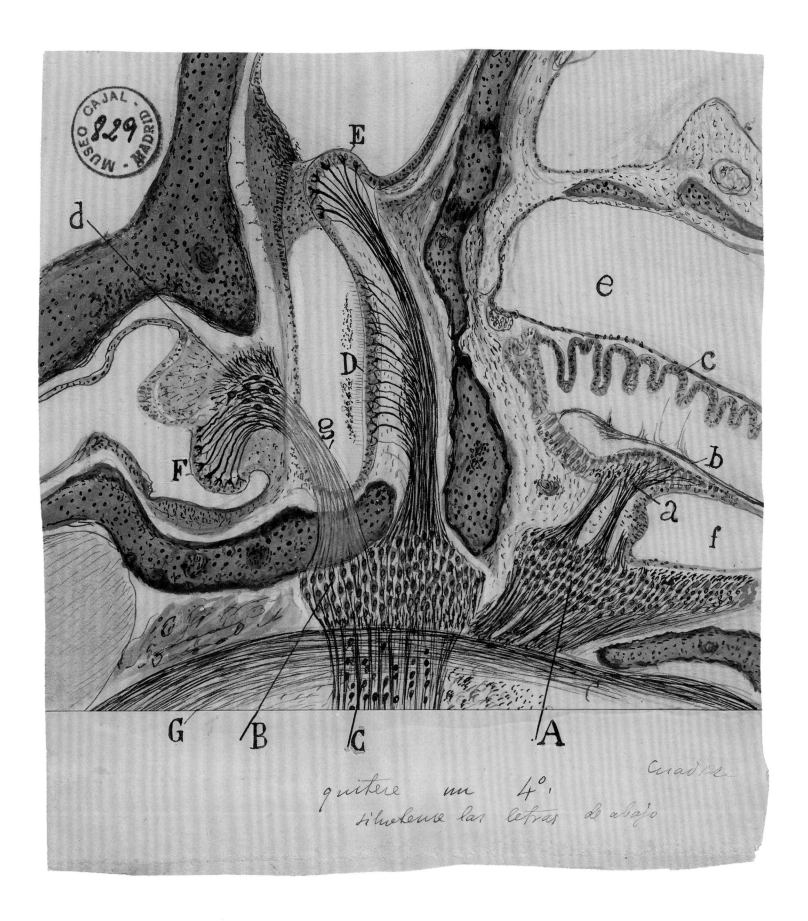

The labyrinth of the inner ear

Although Cajal was fascinated by vision and the structure of the retina, he also studied the other sensory organs of the body. Pictured here is the labyrinth of the inner ear, which contains the sensory structures for hearing and balance. A labyrinth is, of course, a maze, and sound waves travel through this maze as they are transformed into electrical signals. The drawing illustrates a slice through this maze. The organ of Corti (see overleaf), which converts sounds into electrical signals, is shown in a and b. The semicircular canals, the structures that detect rotation of the head, are shown in E and F, while the otolith organ, the structure that senses movement and tilting of the head, is shown in D. The neurons that transmit this information to the brain are indicated by A and B.

The organ of Corti in the human inner ear

The business end of the inner ear, where sounds are transformed into electrical signals, is the organ of Corti. Cajal's drawing of the organ of Corti summarizes the critical elements of this structure. Sounds transmitted from the outer ear and the middle ear enter the inner ear and cause tiny hairs (cilia) at the top of inner hair cells (G) to vibrate. These vibrations generate electrical signals within the inner hair cells that then travel down the cochlear nerve (N) to the brain. Additional outer hair cells (C) function as motors and generate movements within the organ of Corti that amplify the signals detected by the inner hair cells.

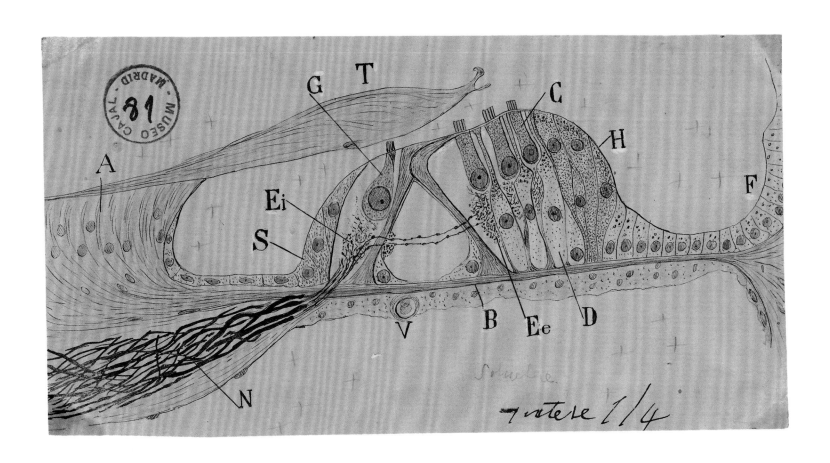

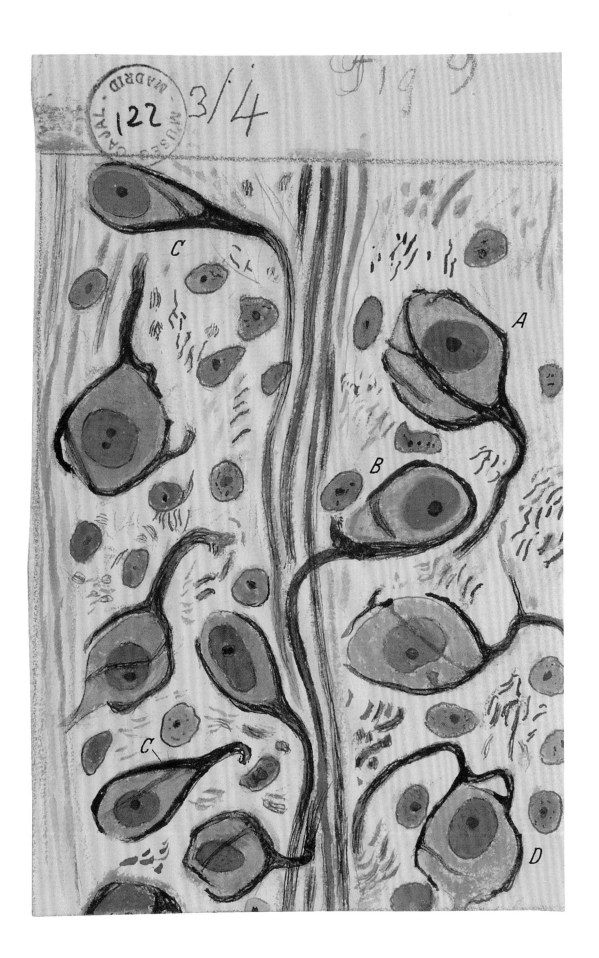

Calyces of Held in the nucleus of the trapezoid body

The calyces of Held are synapses made by axons carrying auditory information and contacting neurons in a brainstem structure called the trapezoid body. The calyces, named by Hans Held in 1883 for their resemblance to calyces of flowers (which envelop the base of flower petals), are the largest synapses in the brain. They are seen in the drawings opposite and on page 31 as stout black lines wrapped around the yellow cells. As was well known to Cajal, these cells are part of the brain system that perceives sound. The large synapses, which transmit information quickly and reliably, help us to accurately localize the source of sounds. Cajal used the structure of this synapse to support his Neuron Doctrine of distinct neurons, stating, "one receives the impression that the chalice is something external to the cell although very intimately applied to it."[9]

Neurons of a child's auditory cortex

Auditory signals originating in the inner ear, after passing through the calyces of Held in the nucleus of the trapezoid body (see pages 31 and 104) and several other brain structures, make their way to the cerebral cortex, where they are perceived as sounds. This drawing shows the structure of one type of neuron that helps process the auditory signals in the auditory region of the cerebral cortex.

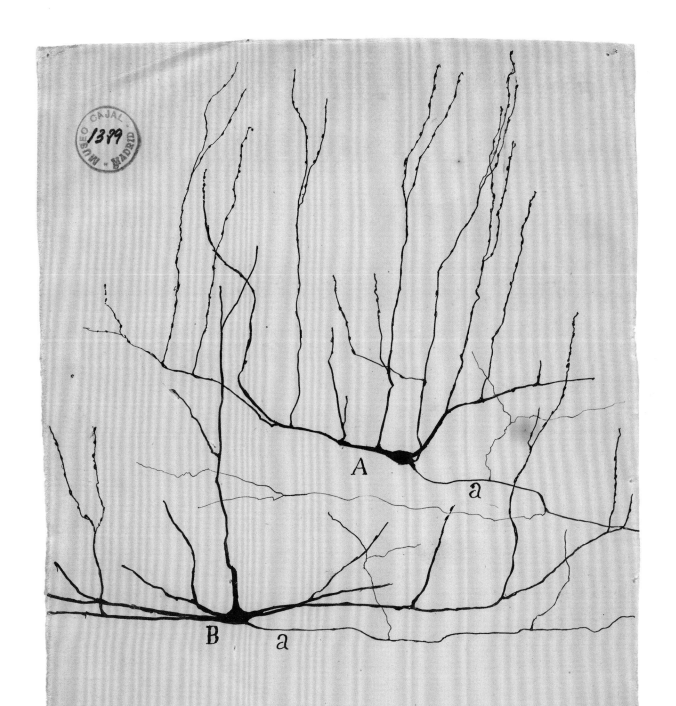

D E G M F A B

Ending of the vestibular nerve

The vestibular nerve transmits information from structures in the inner ear that detect head orientation and head motion. The brain receives this information and uses it to keep our bodies upright when we are standing and to keep our eyes fixed on an object as our head moves. This drawing shows the axons that come from the inner ear as they end in a structure in the brainstem called the vestibular nucleus.

Olfactory pathways within the brain of the rodent

Olfactory information from the nose first enters the brain at the olfactory bulb (A). Olfactory signals then travel directly to the olfactory cortex, where the signals are interpreted as odors. Our sense of smell is, in evolutionary terms, one of our most primitive senses. The region of the cerebral cortex that is devoted to smell is similarly primitive, having a simpler structure (three layers) than regions of the cortex devoted to our other senses (six layers). The pathways of the olfactory system are shown in this drawing, which is a rough sketch, with Cajal's original pencil marks still showing.

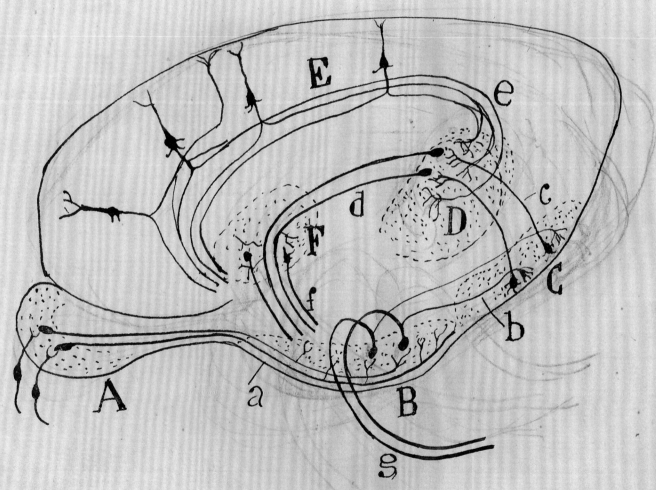

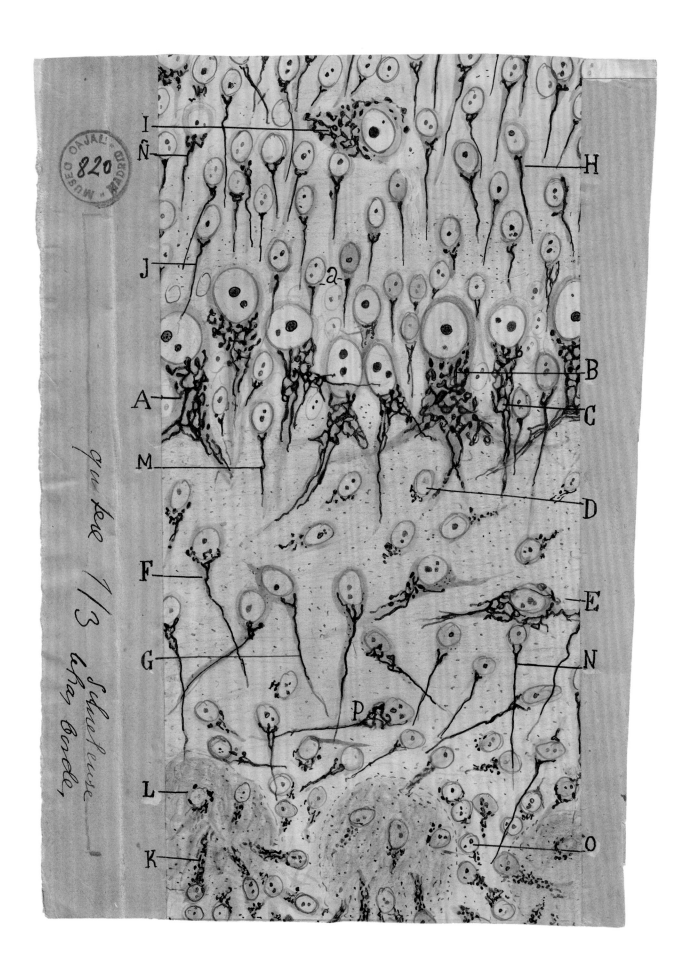

The olfactory bulb of the dog

The olfactory bulb is the brain structure that receives signals coming from the nose. These signals are then sent from the bulb directly to the olfactory cortex, where odors are perceived. In humans, the olfactory bulb lies beneath the main body of the cerebral cortex, just above the nose. The bulb is organized in a systematic manner, with different types of odors exciting neurons in different parts of the bulb. Here, Cajal illustrates the different types of neurons that are present in the olfactory bulb.

The thalamus of the guinea pig

The thalamus is a large structure in the brain that lies between the cerebral cortex and the midbrain. As Cajal describes in his writings, the thalamus functions as a relay station. It receives inputs from our sense organs and sends this information to different areas of the cerebral cortex. In this drawing, Cajal illustrates the structures within the thalamus that handle different types of sensory information. For instance, information from the eyes enters the thalamus from optic tract axons (A) and goes to the lateral geniculate nucleus (B and C), the area of the thalamus devoted to vision. The structure labeled F is the medial geniculate nucleus, which is dedicated to hearing (see overleaf).

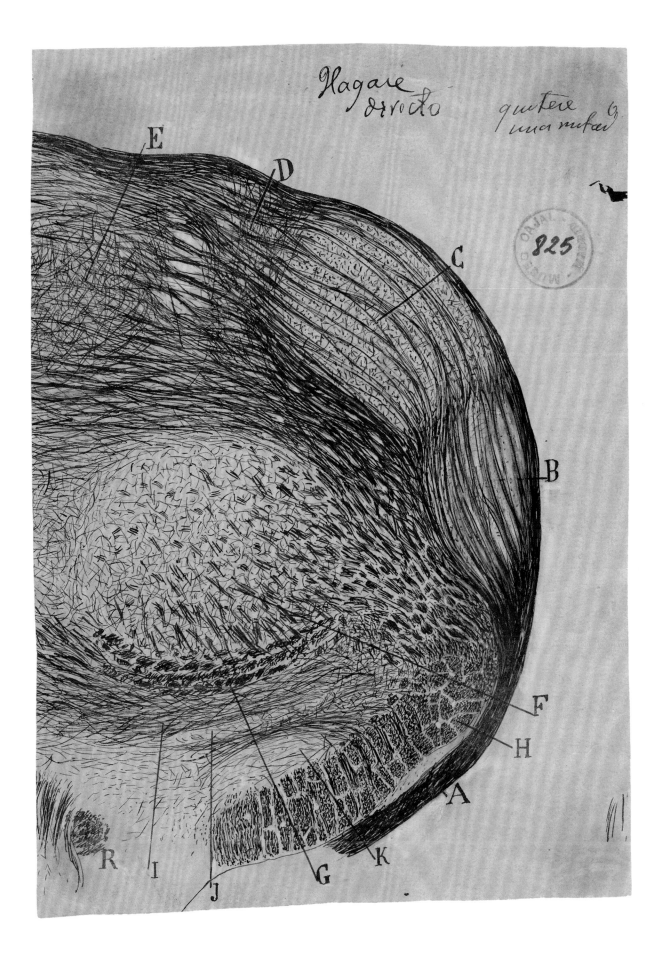

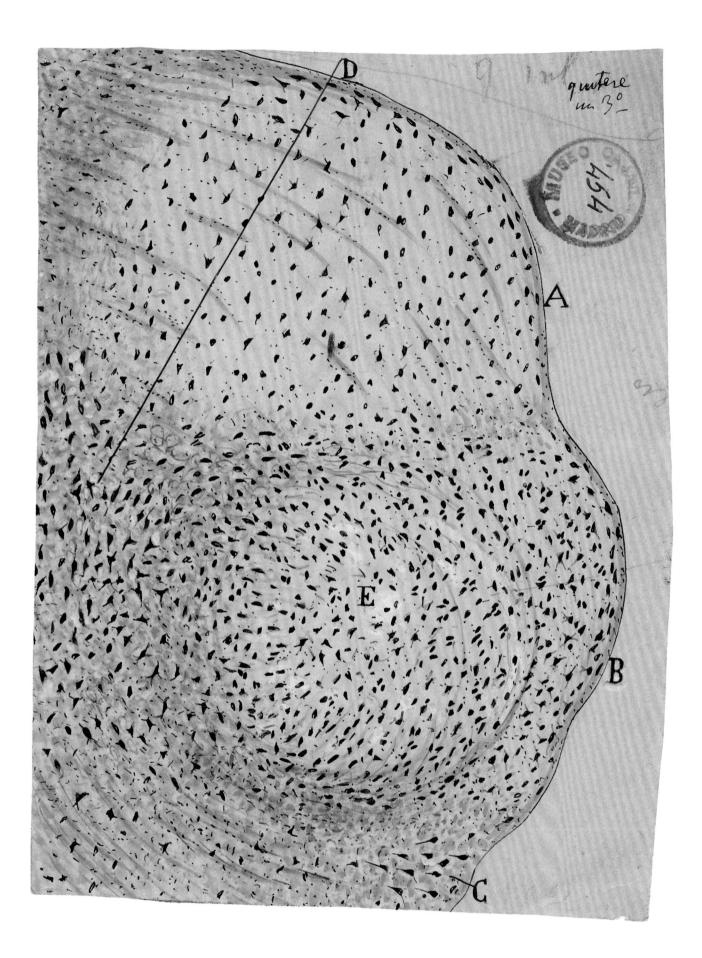

The medial geniculate nucleus in the thalamus of the cat

The medial geniculate body is a structure within the thalamus that deals with hearing. It receives inputs from auditory structures in the midbrain and sends this auditory information on to the cerebral cortex. Like other structures in the thalamus, the medial geniculate nucleus contains neurons that faithfully relay information from the body's sense organs to the cerebral cortex when a person is awake. Unknown to Cajal, however, this relay function is interrupted when a person sleeps, so that sensory information does not reach the cerebral cortex. Cajal shows that the different divisions of the medial geniculate nucleus have different densities of neurons.

The dorsal root ganglion

Sensory information from our skin, muscles, and other organs enters the spinal cord by way of neurons in the dorsal root ganglia, which lie just outside the spinal cord. These are atypical neurons. Instead of having a branched dendritic tree and an axon emerging from opposite sides of their cell bodies, they have two appendages, one traveling outward to the peripheral organs and the other inward toward the spinal cord. Cajal classified the appendage traveling toward the periphery as a dendrite, as it receives sensory input and conducts electrical signals toward the cell body. The appendage traveling into the spinal cord is a traditional axon, as it conducts signals away from the cell body. In this drawing, Cajal illustrates the shape of these specialized neurons in different animals. He shows that in fish (A) each neuron has two appendages emerging from opposite ends of the cell body. In mammals (B), in contrast, these appendages coalesce into one continuous structure that travels from peripheral organs all the way to the spinal cord. This combined dendrite-axon is connected to the cell body by an additional short appendage. Cajal points out that signals traveling from the skin to the spinal cord follow a shorter and straighter course in mammals.

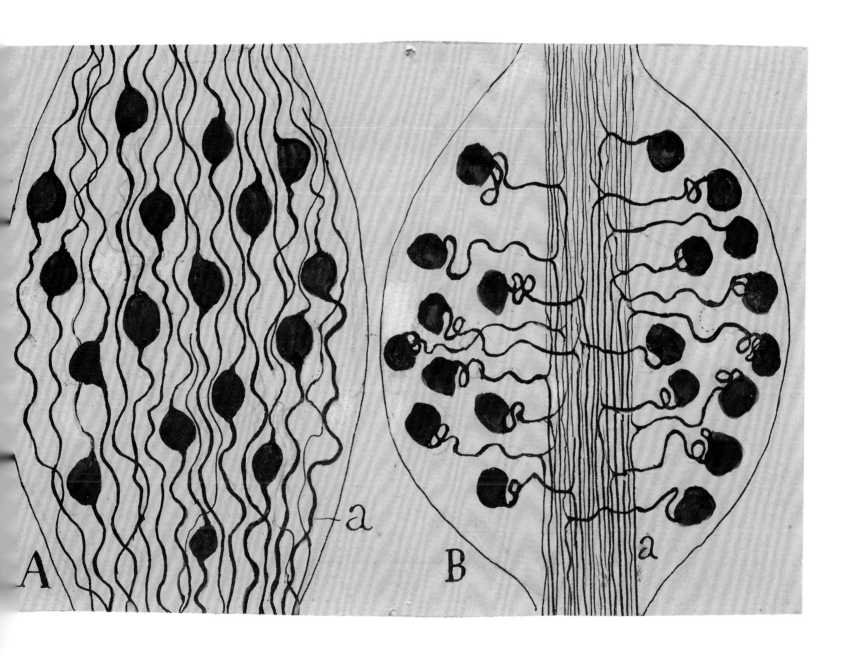

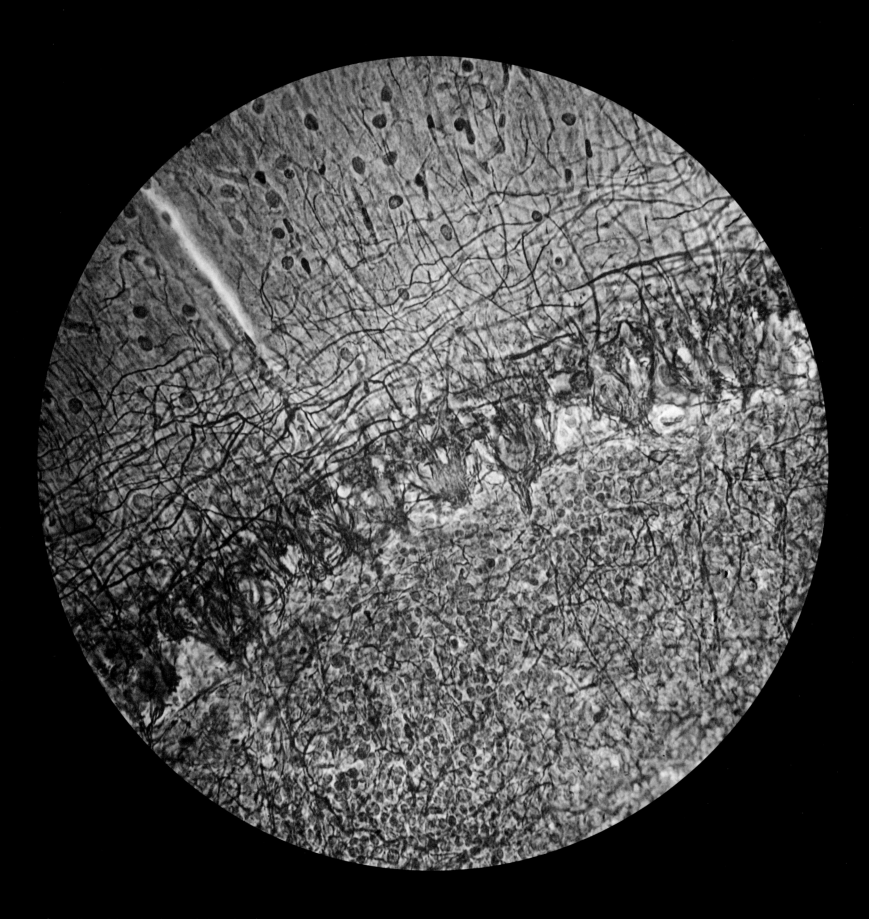

Cerebellum of the cat stained with the reduced silver-nitrate method, microphotograph by Cajal.

NEURONAL **PATHWAYS**

One of Cajal's most important contributions to our understanding of the brain was his discovery of the direction of information flow within neurons and in neuronal circuits. Many of his drawings include arrows that illustrate the direction of information flow within a group of neurons. He made many drawings of the retina that show the direction of information flow in this manner (see, for example, page 94). He also illustrated information flow in local neuronal circuits in the brain, as well as in long-range pathways in the brain and the spinal cord. Some of these drawings are included in the following section.

Cajal could not measure the electrical activity of the cells within the brain. Yet, based on his analysis of the arrangement of the cells in the retina, cerebellum, and olfactory bulb, he postulated the Theory of Dynamic Polarization in 1891. He stated, "The transmission of the nerve impulse is always produced from the dendrites and the cell body to the axon. Every neuron thus has a receiving apparatus, the soma [cell body] and dendrites, an emission apparatus, the axon, and a distributing apparatus, the nerve terminal arborization [branches of the axon making contacts with other neurons]."[10]

Our modern conception of how the brain functions would not be possible without Cajal's Neuron Doctrine and his Theory of Dynamic Polarization. The existence of neuroral pathways and brain structures responsible for specific brain functions is incompatible with the Reticular Theory, which held that information flowed through a continuous interconnected network of cell appendages. But, as Cajal observed, specialized brain pathways subserving specific functions are easily appreciated when it is understood that neurons make specific contacts with other neurons and that information flows through these neuron pathways in a specific direction.

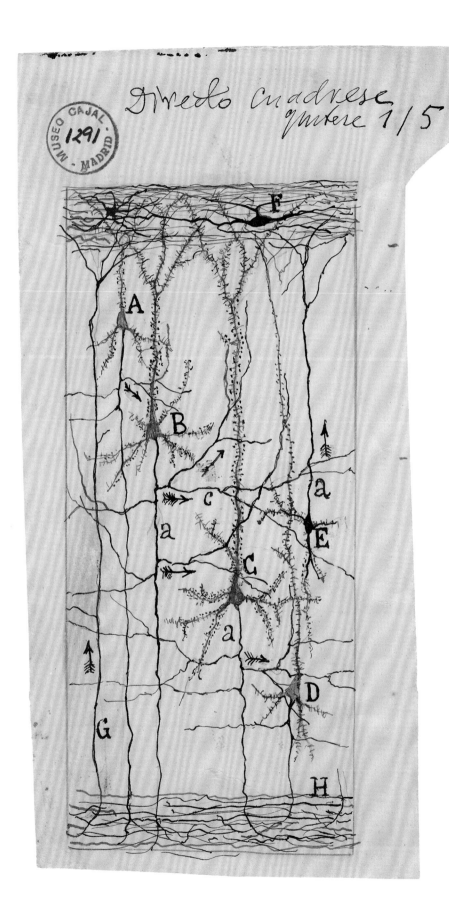

Neurons in the cerebral cortex

The famous British neuroscientist Sir Charles Sherrington, who won the Nobel Prize in 1932, wrote of Cajal, "He solved at a single stroke the great question of the direction of nerve-currents in their travel through the brain and spinal cord. He showed, for instance, that each nerve-path is always a line of one-way traffic only, and that the direction of traffic is at all times irreversibly the same."[11] Cajal illustrates the direction of information flow in this drawing of different types of neurons in the cerebral cortex. Arrows in the drawing indicate the direction of information flow, including signals in axons originating in other brain areas (G), which then make contacts onto local neurons.

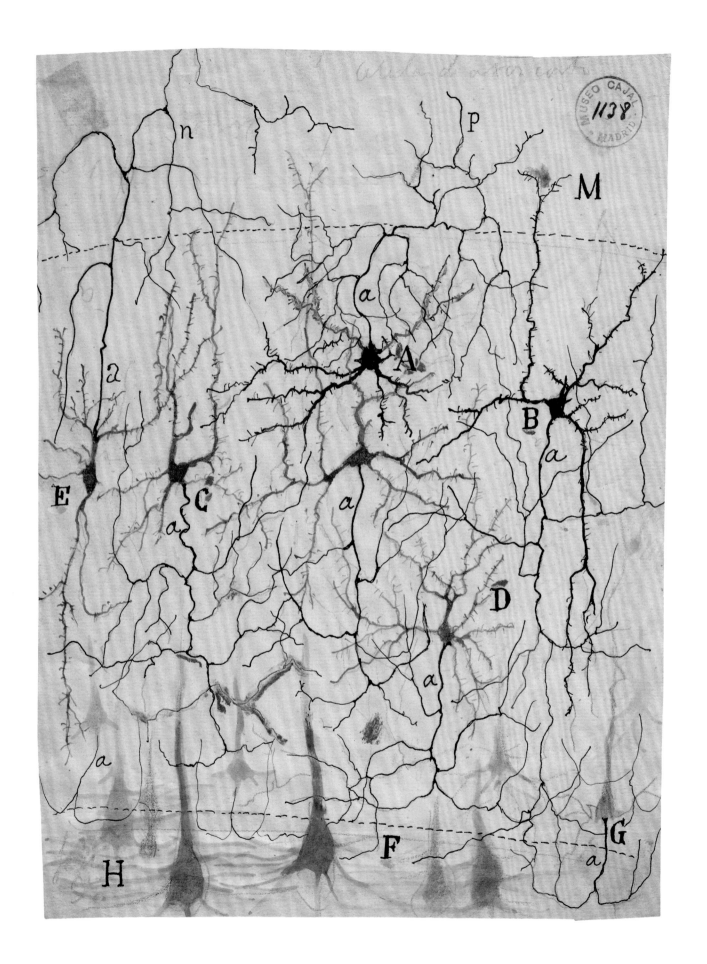

The visual cortex of the cat

Visual information coming from the eyes first travels to a structure deep within the brain called the thalamus (see pages 115 and 116), which then relays the signals to the visual area of the cerebral cortex. Cajal illustrates one specialized type of neuron in the visual cortex called the stellate cell (A, B, C, and E, colored darkly). Stellate cells receive visual inputs directly from the thalamus and then signal other cortical neurons. Processing of this visual information results in our perception of objects. Several pyramidal neurons (colored lightly) are illustrated at the bottom.

Neurons in the middle layers of the cerebral cortex

Cajal's appreciation of the beauty of the brain is expressed in this passage from his autobiography: "Like the entomologist in pursuit of brightly colored butterflies, my attention hunted, in the flower garden of the gray matter [the cerebral cortex], cells with delicate and elegant forms, the mysterious butterflies of the soul, the beating of whose wings may someday—who knows?—clarify the secret of mental life."[12] Cajal illustrates some of his butterflies in this drawing of the cerebral cortex.

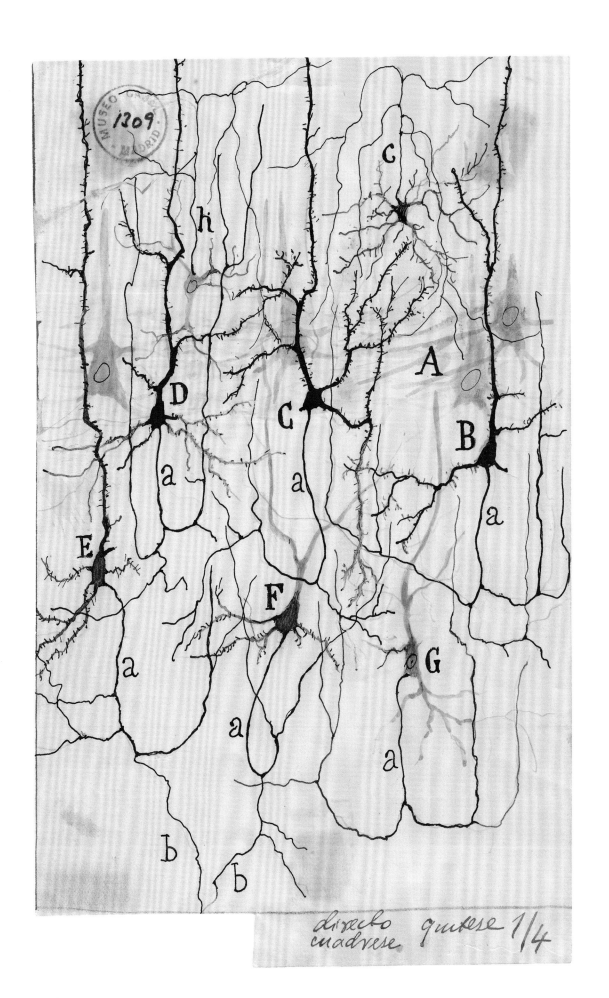

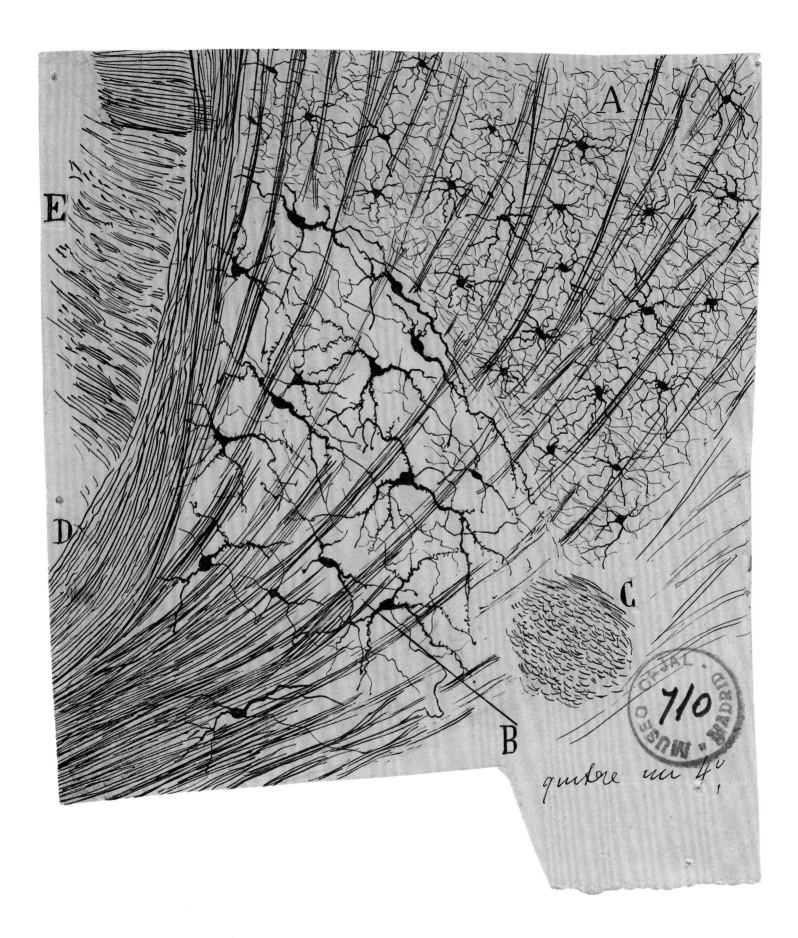

Caudate nucleus of a twenty-day-old mouse

The caudate nucleus is part of a larger structure, the basal ganglia, that lies deep within the brain. This structure plays an essential role in helping us control the movements of our body. The caudate nucleus receives signals from the cerebral cortex that help us coordinate our movements. In Huntington's disease, a devastating neurodegenerative disease, many of the neurons in the caudate nucleus die. This results in uncoordinated, jerky body movements, a hallmark of the disorder. In this drawing, Cajal illustrates neurons in the caudate nucleus (A), along with several fiber tracts (bundles of axons; C, D, and E) that pass through this brain region and transmit information from one brain area to another.

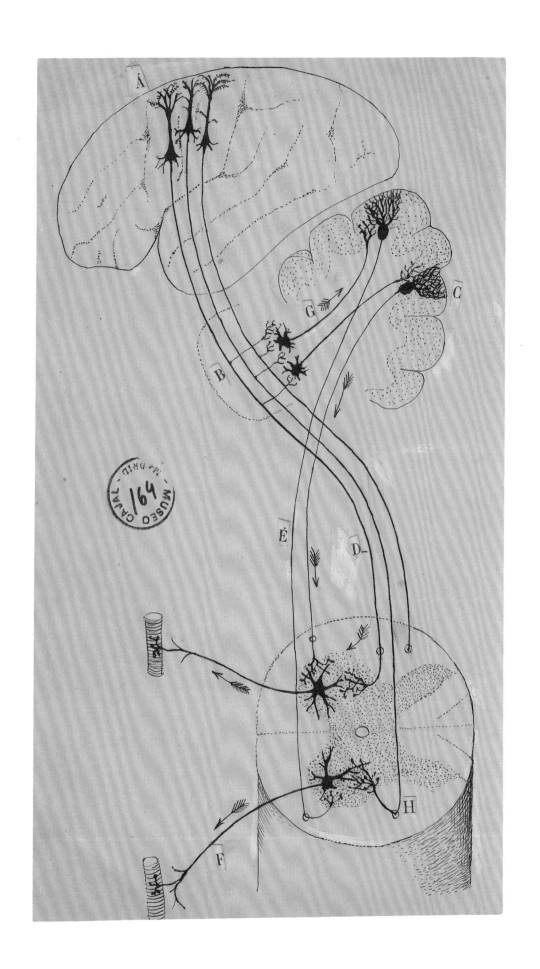

Two pathways that control our muscles

Cajal summarizes the theory that our muscles are controlled by two complementary motor pathways in the brain and spinal cord. In the first pathway, pyramidal neurons in the motor area of the cerebral cortex (A) send signals (D) directly to large neurons in the spinal cord (H) called motor neurons. These motor neurons, in turn, send command signals (F) to our muscles (the two tube-shaped cells at the lower left). In the second motor pathway, Cajal shows that Purkinje neurons in the cerebellum (C) also send signals (E) to the spinal cord that indirectly influence the motor neurons. Cajal discusses in his writings how these two pathways might interact. For instance, signals from the cerebral cortex might also influence the Purkinje neurons of the cerebellum by way of pathway G.

Sensory pathways within the spinal cord

Cajal illustrates in this drawing the pathways of sensory signals as they enter the spinal cord. Sensory signals from our skin and muscles are transmitted to the spinal cord by neurons in the dorsal root ganglia (DL and C). (The drawing on page 119 shows the neurons of a dorsal root ganglion in greater detail.) When the signals enter the spinal cord, they are transmitted up to structures in the brainstem (d, e, and f) by way of axons (b and c). Sensory signals from the lower body enter the spinal cord through thoracic and lumbar dorsal roots (DL) and end in the gracile nucleus (d), while signals from the upper body enter the spinal cord through cervical dorsal roots (C) and end in the cuneate nucleus (e). (The drawing on page 152 shows the cuneate nucleus in detail.) Sensory signals also travel within the spinal cord through branches of the dorsal root ganglion axons (a). These signals are involved in generating our motor reflexes.

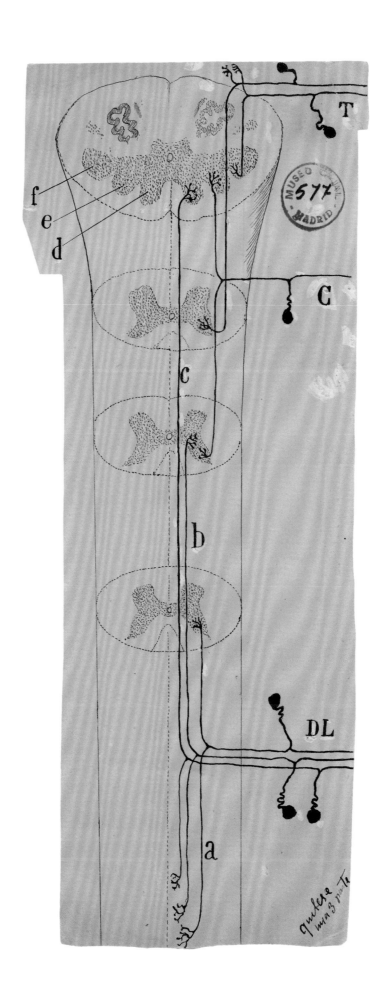

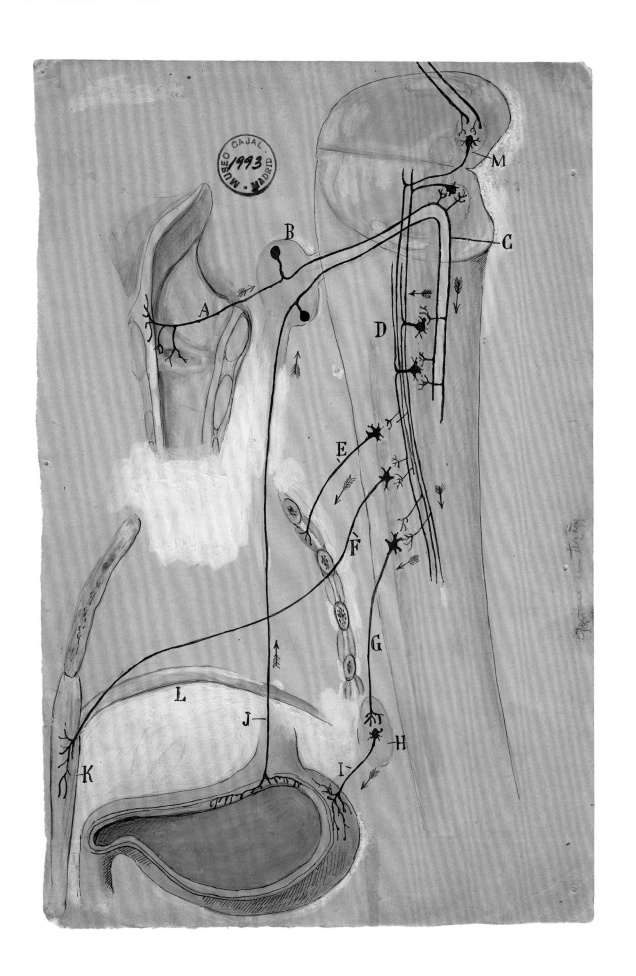

Pathways mediating the vomiting and coughing reflexes

In addition to documenting the structure of the brain, Cajal outlined the brain pathways that control different behaviors. This drawing illustrates the signaling pathways responsible for the vomiting and coughing reflexes. When the larynx in the throat (A) is irritated, signals are sent via the vagus nerve (B) to the brainstem (M) and spinal cord (D). These signals excite neurons in the spinal cord that cause muscles in the chest and abdomen (K) to contract, causing us to cough. When the lining of the stomach (at bottom) is irritated, signals are sent through another branch of the vagus nerve (J) to the spinal cord. This excites a neuronal pathway (G, H, and I) that causes the stomach to contract and us to vomit. Modern research has shown that vomiting can also be triggered by harmful chemicals in the blood activating neurons in the brain.

The structure and connections of the hippocampus

The hippocampus is, in evolutionary terms, a primitive part of the cerebral cortex that plays an essential role in the consolidation of our memories. The structure is named after its striking resemblance to a seahorse (whose Greek name is *hippocampus*). Cajal and his contemporaries referred to the main body of the hippocampus as Ammon's horn, for its resemblance to the ram's horns of the Egyptian god Ammon. The structure of the hippocampus is simpler than other parts of the cerebral cortex (having three cell layers instead of six), but, like the neocortex, its most prominent neurons are pyramidal neurons. These neurons are present throughout the hippocampus (c, h, and g). Cajal indicates the flow of information within the hippocampus with arrows.

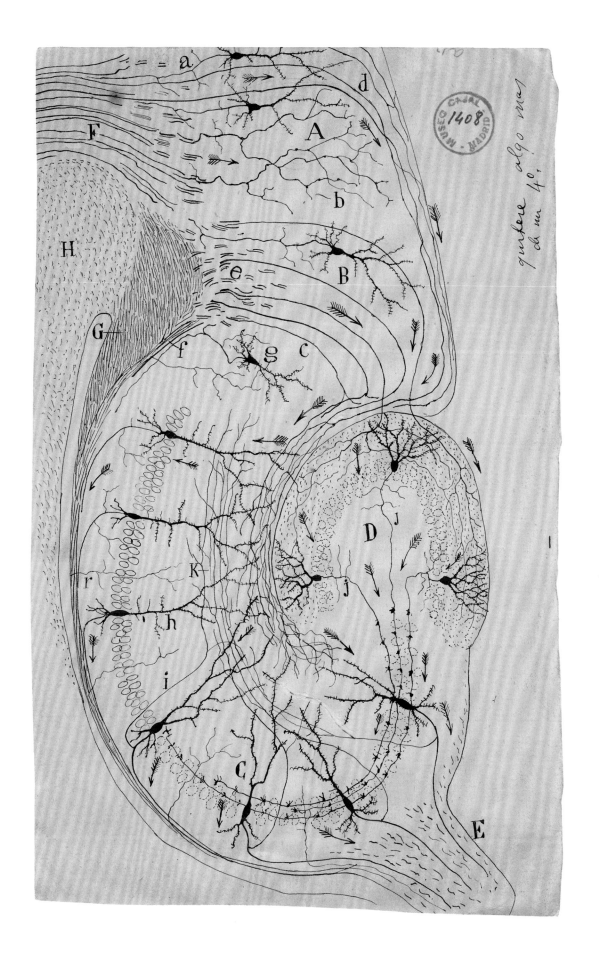

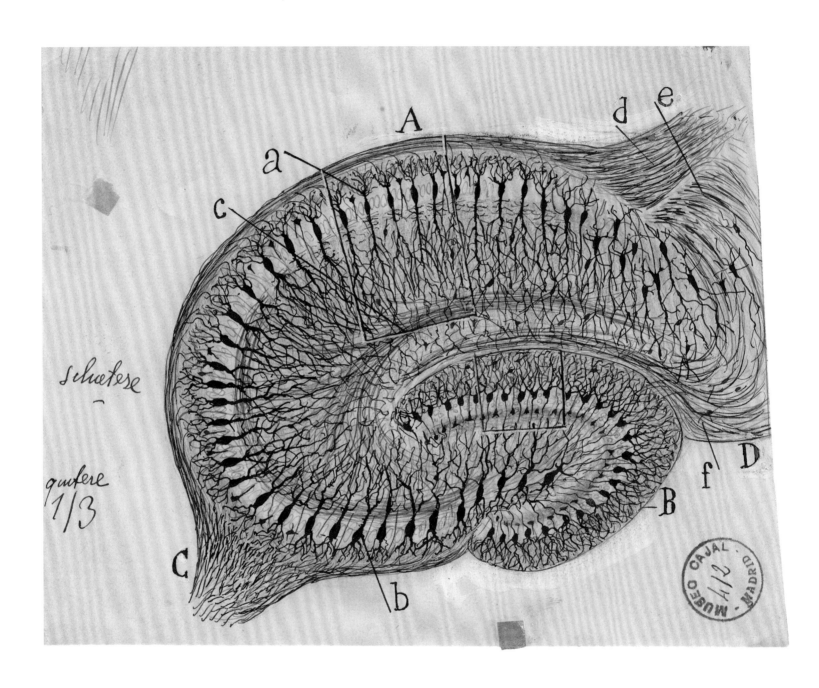

The hippocampus

Cajal highlights the prominent role of pyramidal neurons in the hippocampus in this drawing. The pyramidal neurons are the darkly colored cells (a and b) whose cell bodies lie near the outside of the hippocampus and whose dendrites extend toward the center. Cajal waxes poetic about hippocampal pyramidal neurons in his autobiography: "[The] pyramidal cells, like the plants in a garden—as it were, a series of hyacinths—are lined up in hedges which describe graceful curves."[13]

Connections within the hippocampus

This drawing illustrates connections between different areas of the hippocampus, with axons (B) from the dentate gyrus (A) contacting and forming synapses onto pyramidal neurons (C) in Ammon's horn. Cajal was well aware of the importance of the hippocampus in memory formation, stating in his autobiography that the hippocampus was "the oldest center of association in the brain, the storehouse of olfactory memories."[14] What Cajal could not know, but what modern research has revealed, is that the hippocampus is absolutely essential for the formation of new memories. If the hippocampi on both sides of the brain are destroyed, a person is incapable of forming new memories.

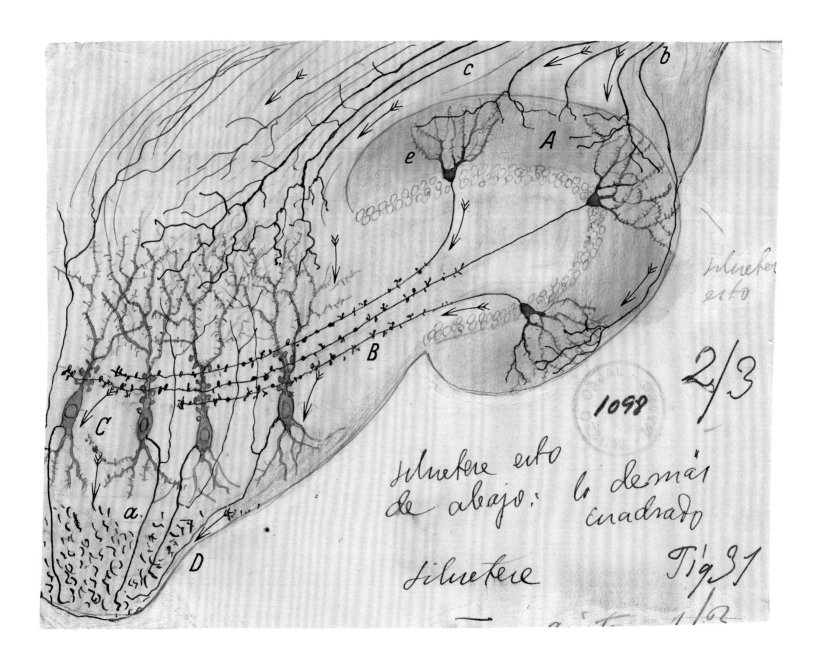

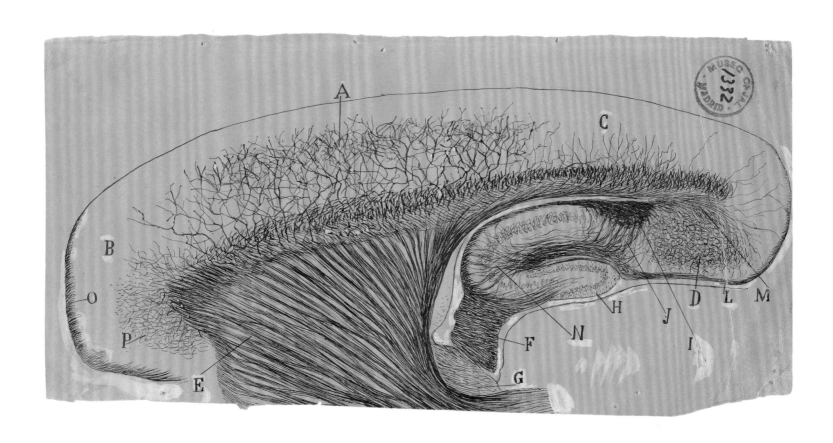

Structures deep within the brain

Cajal illustrates several of the structures deep within the brain, including the internal capsule (E), which contains bundles of axons that connect the cerebral cortex (B and C) with other parts of the brain. Label A shows the endings of some of these axons in the cortex. Also illustrated is the dentate gyrus (H) and Ammon's horn (N), the two interlocking parts of the hippocampus. The dentate gyrus contributes to the formation of our memories and is one of the few places where new neurons are born in the adult brain. Recent research suggests that the birth of these neurons in the dentate gyrus is essential and that a decline in the generation of new neurons is associated with depression.

Ventral surface of the medulla and pons in the human brain

This is an atypical drawing by Cajal, as it shows the three-dimensional shape of the surface of the brain rather than the structures and individual cells revealed in thin slices of the brain. The picture is of the brainstem, as viewed from a person's front. Cajal highlights the cranial nerves (II through XII), the bundles of axons that connect the brain primarily to regions of the head and the neck. For instance, the optic tract (II) connects the brain to the eyes. The crossing of about half of the optic tract axons from one side of the body to the other is easily seen (see also page 87). Other cranial nerves include the trigeminal nerve (V) and the vagus nerve (X).

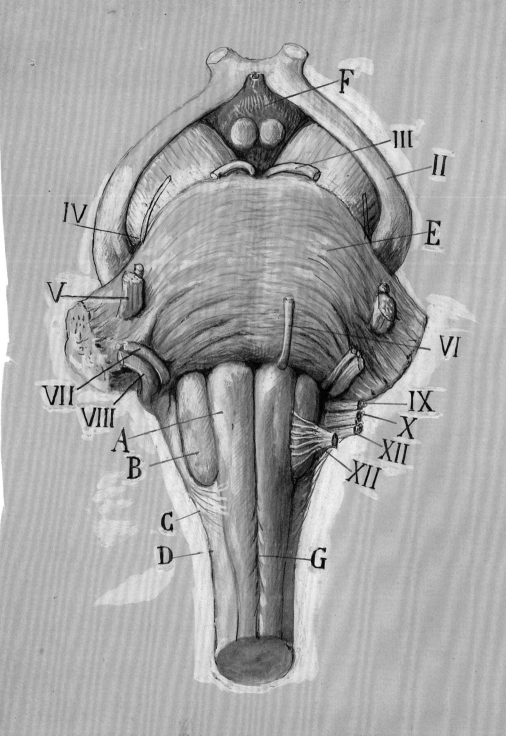

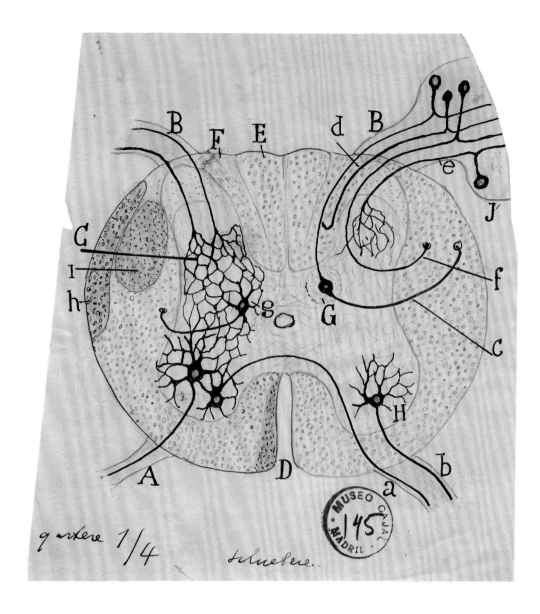

Contrasting theories of the composition of the brain

Cajal illustrates contested theories of the composition of the brain in these two drawings of the spinal cord. The Reticular Theory of the brain held sway early in Cajal's career. This theory, illustrated in the left-hand drawing, proposed that the brain was composed of a network of continuously connected cells. Neuron cell bodies (g) are joined by an interconnected network of cell appendages (C). Cajal, in contrast, correctly observed that the brain was composed of discrete neurons that were separated by gaps (right-hand drawing; k, l, m, n, s, t, u). Cajal won over his colleagues to his Neuron Doctrine, which was conclusively confirmed in the 1950s.

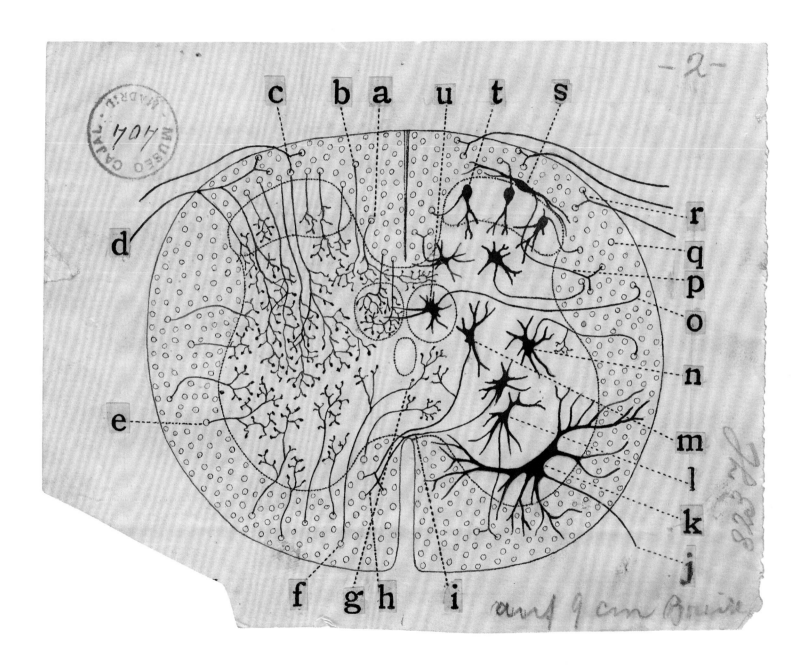

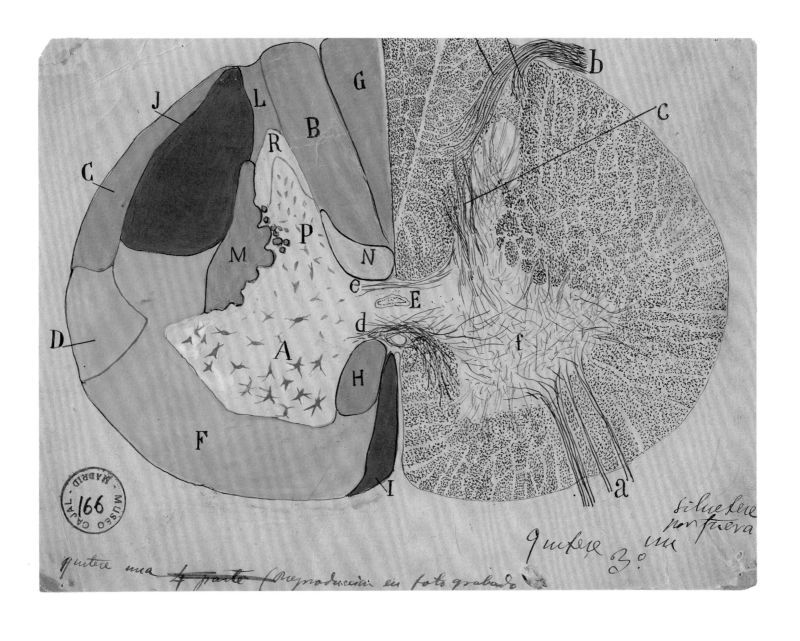

Axon fiber tracts in the spinal cord

Long-distance signaling pathways are needed for the brain to properly control the body. Signals from the motor centers of the brain must be sent to our muscles to make them contract. Signals from our muscles, joints, and skin must also be sent back to the brain so that it knows what our body is doing. Many of these signals travel in bundles of axons within fiber tracts in the spinal cord. Cajal illustrates these fiber tracts schematically on the left side of this drawing of the spinal cord. For instance, signals from the motor centers of the cerebral cortex travel down the spinal cord through the lateral corticospinal tract (J), while information from our skin signaling touch and vibration travels up to the brain through the dorsal column tracts (B and G). The actual appearance of the axon bundles is shown on the right side of the drawing.

The embryonic spinal cord

The spinal cord has many fiber tracts composed of bundles of axons that transmit information between our body and our brain. These fiber tracts are illustrated in this drawing by areas of stippling around the edges of the spinal cord and are shown in greater detail on page 148. Neurons within a single level or segment of the spinal cord are also extensively interconnected. Cajal highlights these axon connections in this drawing. Prominently displayed are the axons of the dorsal root ganglion (D and E) that transmit sensory information from our muscles and skin into the spinal cord. Axon connections (A and B) transmitting information between the two sides of the spinal cord are also shown. These crossed connections help coordinate our movements. For instance, when we bend sideways, muscles on one side of our body contract while muscles on the other side relax.

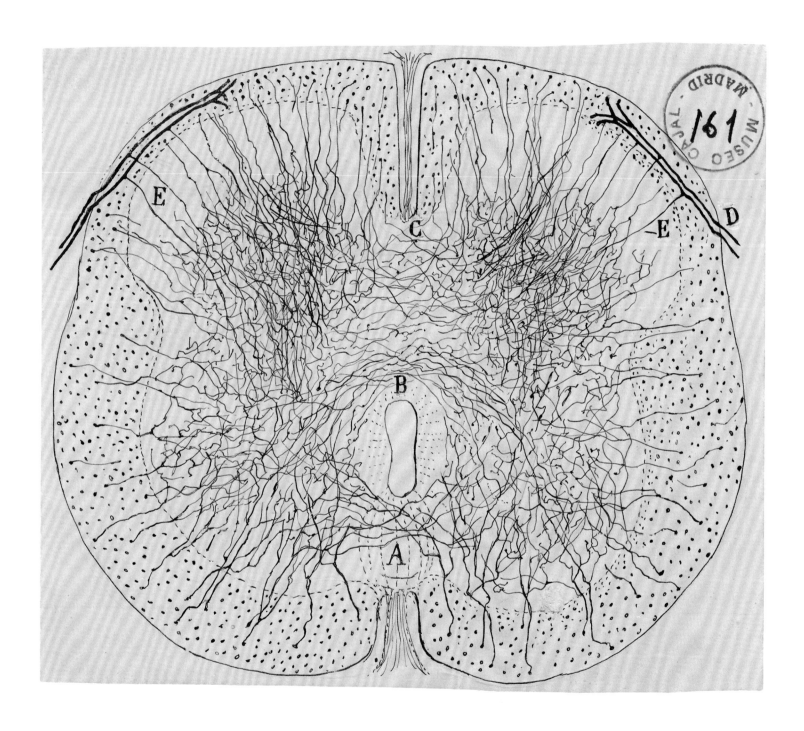

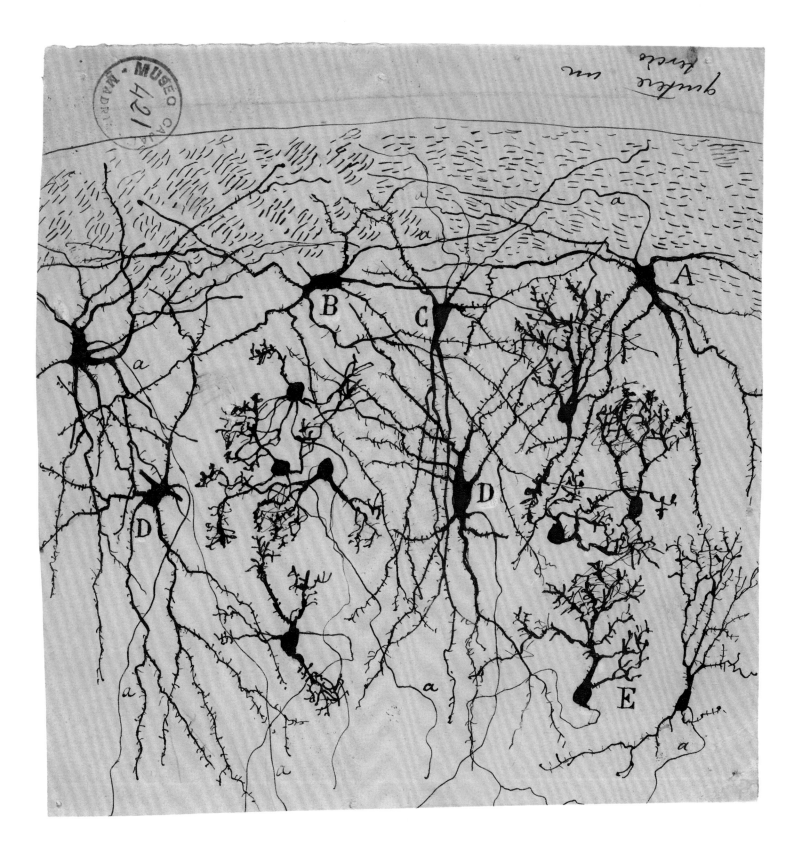

Cuneate nucleus of the kitten

Sensory signals from the skin and the muscles of the hands and arms travel up through the spinal cord and into a wedge-shaped structure called the cuneate nucleus, which lies at the base of the brain just above the spinal cord (see page 133). This structure then relays the sensory information to higher brain areas. Here, Cajal illustrates the different types of neurons within the cuneate nucleus. Many of the neurons have dendrites that are covered with spines, the appendages that receive synaptic contacts from other neurons.

Neurons in the midbrain of a sixteen-day-old trout

There are several clusters of neurons (nuclei) in the midbrain that control our eye and head movements. Two of these neuron groups are shown in this drawing (A and B). Cell group B is the nucleus of the oculomotor nerve. The axons of these neurons (E) form the third cranial nerve, which controls eye movements. (The cranial nerves are shown in the drawing on page 145.) The group of cells labeled A is named in Cajal's honor, the interstitial nucleus of Cajal. Recent research indicates that these neurons help to control vertical movements of our eyes.

quítese un tercio
o sea algo más
en 2/3 quítese la ~~mitad~~
Directo

Silueteense las
letras y los
fondos que
llevan cruces

MUSEO CAJAL 1933 MADRID

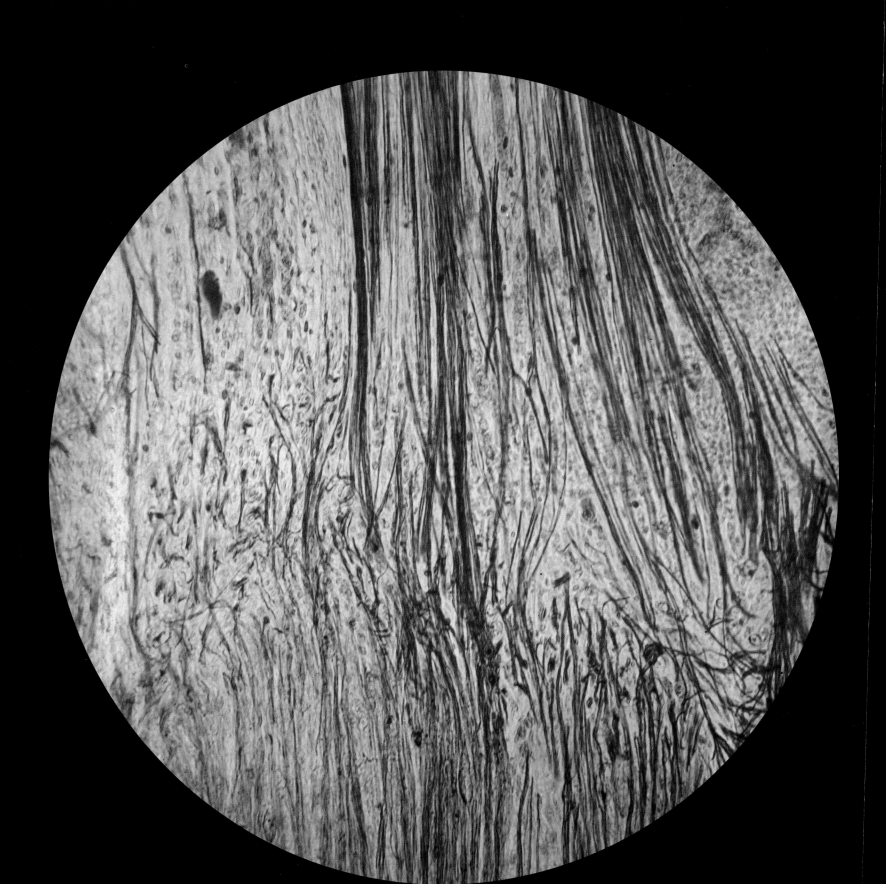

DEVELOPMENT AND PATHOLOGY

Many of Cajal's drawings illustrate neurons or structures from the fully developed adult brain. However, Cajal was also fascinated by the changes that the brain undergoes as it matures, and he studied and documented the brain at many stages of development. The British neuroscientist Sir Charles Sherrington wrote, "A trait very noticeable in [Cajal] was that in describing what the microscope showed, he spoke habitually as though it were a living scene. . . . He treated the microscope scene as if it were alive. . . . A nerve-cell by its emergent fiber groped to find another!"[15]

Cajal conveyed this sense of the living brain by using sequences of illustrations in his drawings, showing progression or change of a cell over time. To show how a neuron matured in the developing brain, for instance, he drew the neuron several times, each with a longer axon (opposite).

Cajal used his observations of the developing brains of embryos and young animals, which have a simpler structure than adult brains, to test his theory that the brain is composed of discrete neurons. This strategy proved to be of paramount importance in verifying the Neuron Doctrine.

Cajal was also intrigued by the changes that occurred in the brain following injury. He was not only a passive observer of the brain but he would also generate injuries in specific brain areas of an animal to determine how the brain responded. For instance, he would cut a peripheral nerve containing a bundle of axons in order to observe how those axons degenerated and whether they repaired themselves (see page 179).

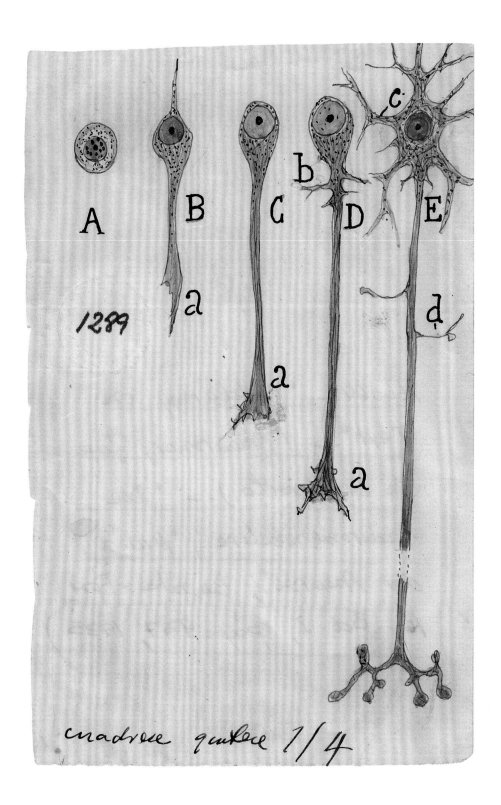

Stages of neuron development

In order to illustrate how a neuron matured, Cajal drew a series of five images of the neuron (A through E) at different stages of its development. As the neuron developed, an appendage, the axon (a), grew out from the cell body. The leading edge of the axon, which guides the axon to its target, is called the growth cone. In the mature neuron (E), the growth cone has found its target, and the axon has formed synaptic contacts, the small circles at the ends of the axon branches. Cajal, using special staining methods, was the first to observe the leading end of the growing axon, and he called it the growth cone, the name we use to this day. The structure of the growth cone is shown in greater detail on page 161.

DEVELOPMENT AND PATHOLOGY

Growth cones in the spinal cord of a chicken embryo

Cajal's discovery of the growth cone at the leading end of growing axons supported his Neuron Doctrine, which held that the brain was composed of discrete neurons. In his autobiography, Cajal described the growth cone: "[T]his ending [of the growing axon] appeared as a concentration of protoplasm of conical form, endowed with amoeboid movements. It could be compared to a living battering-ram, soft and flexible, which advances, pushing aside mechanically the obstacles which it finds in its way, until it reaches the area of its peripheral distribution."[16]

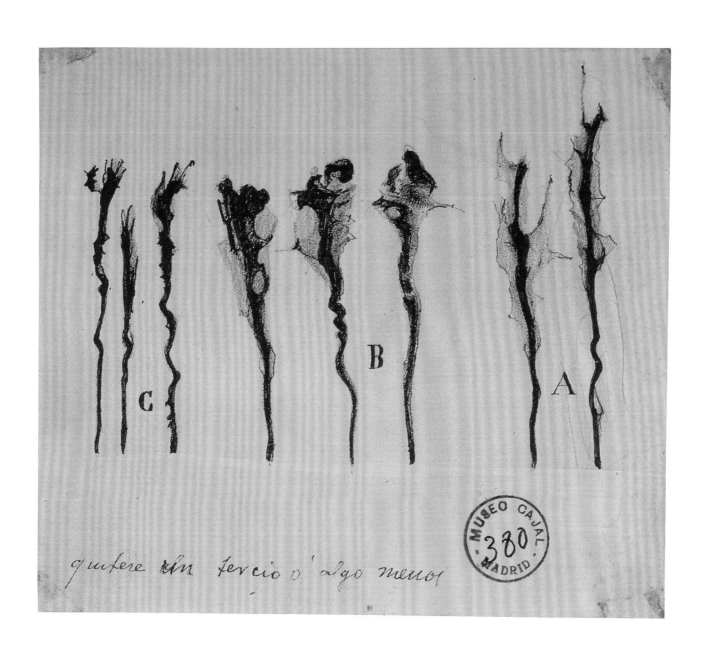

quitese un tercio ó algo menos

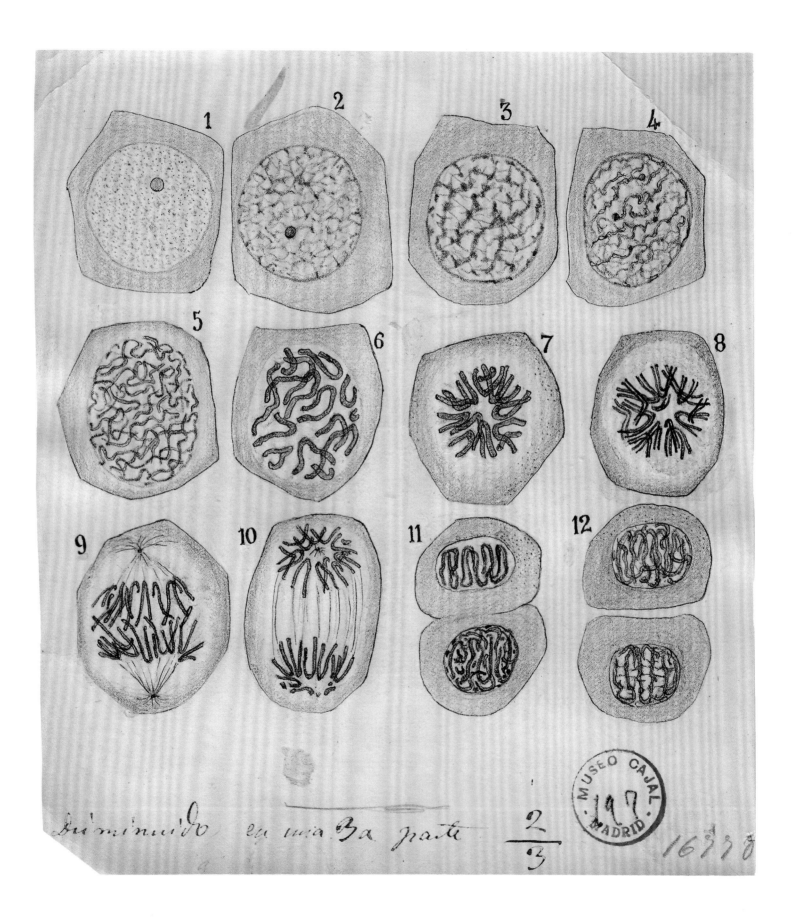

disminuido en una 3a parte 2/3

Division of a skin cell

Cajal had many interests besides the structure of the brain, particularly in his early career. One of these interests was how cells divide and the role of the nucleus (the structure within the cell that contains the chromosomes) in this process. This was a hot topic during Cajal's day. This drawing illustrates the division of a skin cell in twelve steps. The dark, wormlike structures within the cell are the chromosomes, which divide and condense (2 through 6), organize themselves (7 through 9), and segregate at the two ends of the cell (10) just before the cell divides in two (11). This separation of chromosomes, which contain the cell's genetic material, ensures that the two daughter cells each have a full complement of DNA.

Axons of Purkinje neurons in the cerebellum of a drowned man

Cajal was fascinated by the changes that occurred in the brain following injury or, in this case, death. He conducted many studies on the effects of injury on the cerebellum, the results of which are documented in this and subsequent drawings. Here, the cell bodies of the Purkinje neurons of the cerebellum are shown as large, lightly shaded cells near the top. The axons of the Purkinje neurons, emerging from the bottoms of the cell bodies, are shown in various stages of degeneration. The axons are darkly colored. Some (A and B) have disintegrated completely while others (C, D, F, G, and H) remain intact but show bulbous enlargements, a sure sign of degeneration.

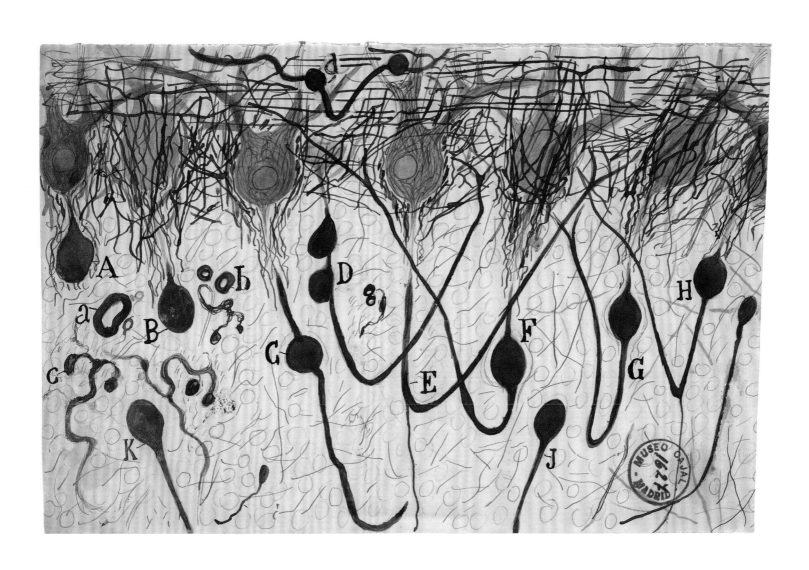

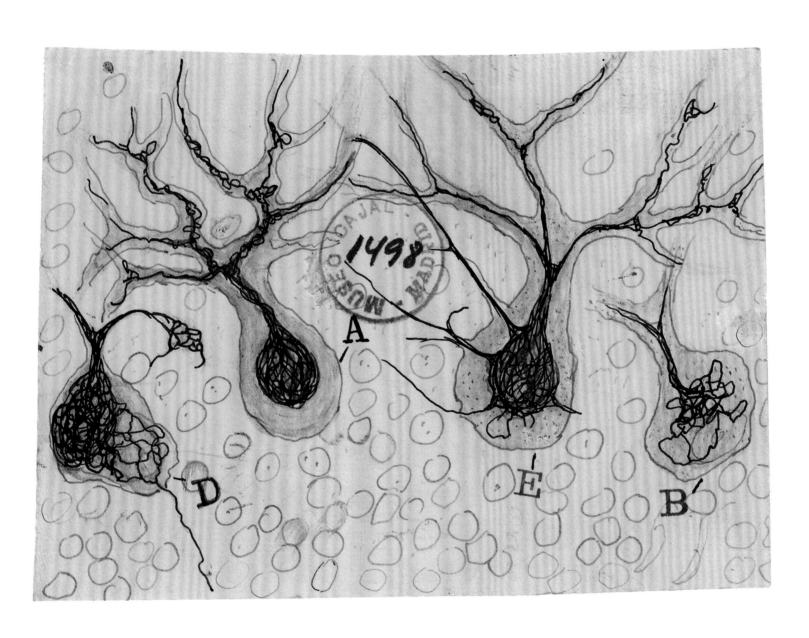

Injured Purkinje neurons

Cajal damaged Purkinje neurons by cutting or compressing nearby tissue. This drawing shows four Purkinje neurons in various stages of degeneration following a compression injury. Cajal noted that following this type of injury, the neurons did not die suddenly, but degenerated gradually over many days, beginning with the outer layers of the cell body. Cajal also found that when the axons of neurons within the brain were cut, the cut ends of the axons died away, but the cell bodies and dendrites of the neurons sometimes survived.

Injured Purkinje neurons in a cat

Cajal conducted a number of studies on damage to Purkinje neurons caused by injury to the cerebellum. Here, Cajal assessed the damage ten days after injury in a cat. The major damage to these neurons occurred in their dendrites, which appear bloated and, in some cases, contain large vacuoles (a). The enlargements and vacuoles seen in these injured cells would be absent in the dendrites of healthy Purkinje neurons (see pages 48–51). Cajal noted that in injured Purkinje neurons, the axon usually degenerated. The Purkinje neurons illustrated in this drawing represent an exception, having healthy axons emerging from the bottom of their cell bodies (A and B).

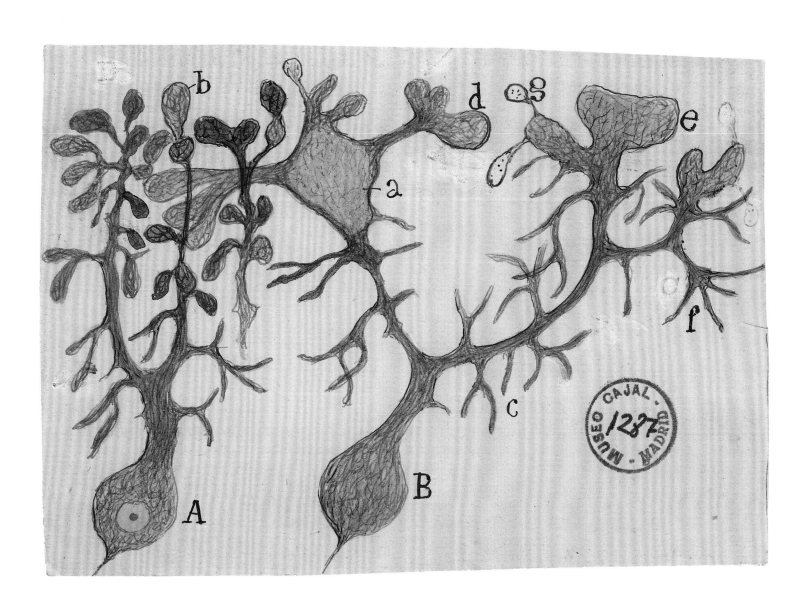

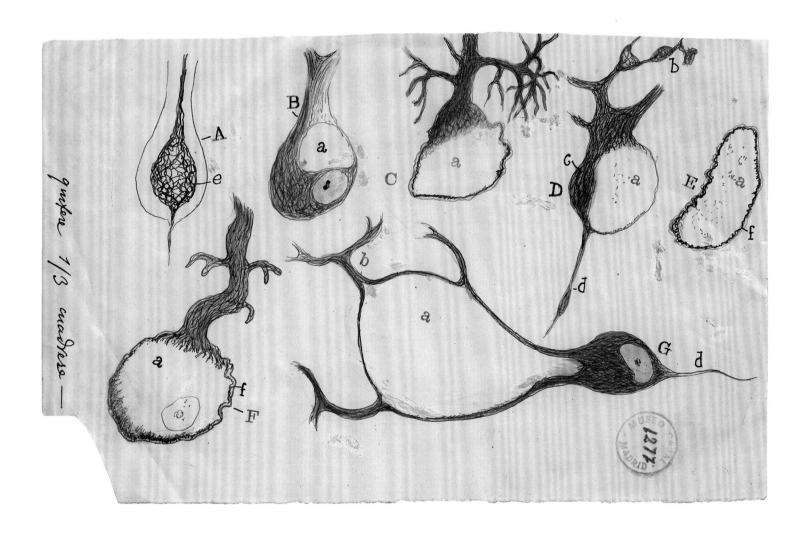

Injured Purkinje neurons of the cerebellum

Judging from this drawing, Cajal must have had a droll sense of humor. Pictured are damaged Purkinje neurons of the cerebellum. Cajal has chosen to focus on the cell bodies, which have a bloated, empty appearance. Large vacuoles (a) are present in several cells, a sure indication of degeneration. The most swollen of these cells (G) has the appearance of a penguin, swimming among the other neurons. Did Cajal actually see a penguin as he gazed into his microscope? Hard to tell, as Cajal typically drew from memory rather than tracing specific cells that he saw.

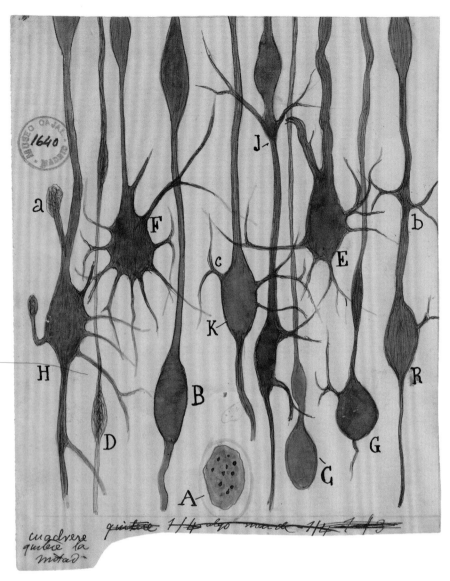

Injured axons of pyramidal neurons in the cerebral cortex

Cajal studied the changes that occur following injury in many parts of the brain. In addition to the extensive investigations of the cerebellum, he characterized the transformation of neurons in the cerebral cortex following tissue damage. These two drawings document changes that take place in the axons of pyramidal neurons. Some axons (A in the left drawing and g and h in the right drawing) degenerate completely, leaving only a ball of decaying tissue, while other axons remain intact but develop large, bulbous varicosities, similar to the ones seen in damaged Purkinje cell axons (see page 165). One of Cajal's main conclusions from these studies was that the brain, unlike peripheral nerves, is unable to regenerate. This lack of regenerative power is what makes damage to the brain and spinal cord so devastating—the brain is unable to repair itself after injury.

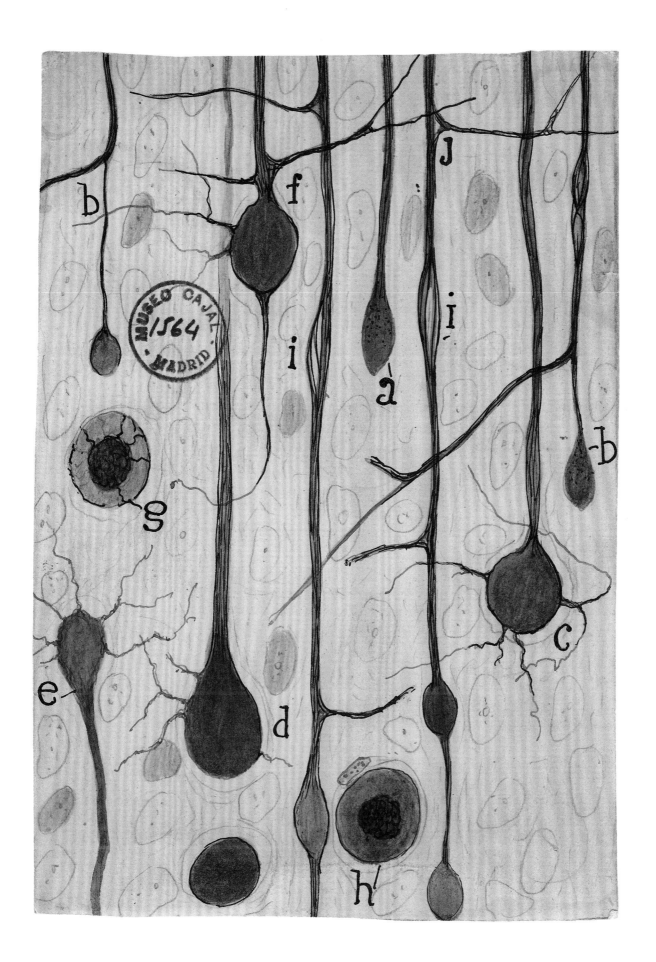

Scar tissue in the cerebral cortex following injury

Following damage to the brain, scar tissue develops in injured areas. This scar tissue is composed primarily of astrocytes, the most common type of glial cell in the brain. The injury causes astrocytes to multiply and grow larger. This drawing shows these transformed astrocytes within scar tissue that formed following injury to the cerebral cortex. Cajal has highlighted the astrocytes by drawing them darkly. Contacts between astrocytes and a blood vessel (the structure oriented diagonally upward from the lower left) can be seen. The cut end of an axon cannot regrow through astrocyte scar tissue. This is one of the reasons why neurons in the brain cannot regenerate following injury and why the brain cannot repair itself. Astrocyte scar tissue can also be the source of epileptic seizures that are generated in the brain following injury.

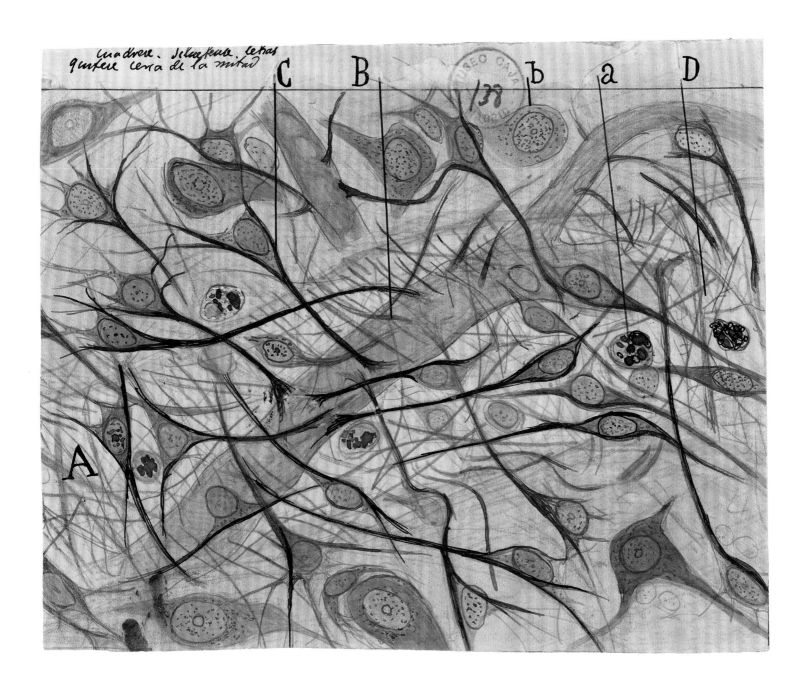

Glial cells in the cerebral cortex of a man who suffered from paralysis

Signs of pathology are often visible in the brains of patients who suffer from injury or neurodegenerative diseases. Amyloid plaques and neurofibrillary tangles, for instance, accumulate in the brains of Alzheimer's patients. Cajal examined the brain of a man who had suffered from paralysis. As illustrated in this drawing, he found transformed glial cells (A) in the cerebral cortex of this patient. These glial cells, which have a dense accumulation of darkly stained fibers, are closely associated with blood vessels (the tubelike structures at the lower left containing red blood cells). Did Cajal include a ghostly face in the upper left corner of the drawing looking down upon the transformed glial cells?

A cut nerve outside the spinal cord

In addition to studying the consequences of injury within the brain, Cajal also investigated the effects of cutting peripheral nerves, the axon bundles that leave the spinal cord to connect with other organs of the body. This drawing shows the cut ends of a nerve after it was severed. The central stump (A), the one still attached to the spinal cord, is shown at the top, while the disconnected end of the nerve (B) is at the bottom. The axons of the central stump are still connected to their cell bodies, so they remain alive and are able to regrow. These newly emerging axons, with their growth cones (the enlargements at the ends of the axons), are shown growing in somewhat random directions from the central stump. This tangle of axons extending from the injured nerve sometimes forms a neuroma, which can be a source of intense pain. Some of the axons growing from the central stump (f and g) have made it across to the disconnected end of the nerve and will regrow through the nerve all the way back to their target organ. Cajal has illustrated a basic principle in this drawing: A peripheral nerve, unlike the nerves of the brain or the spinal cord, is able to regenerate.

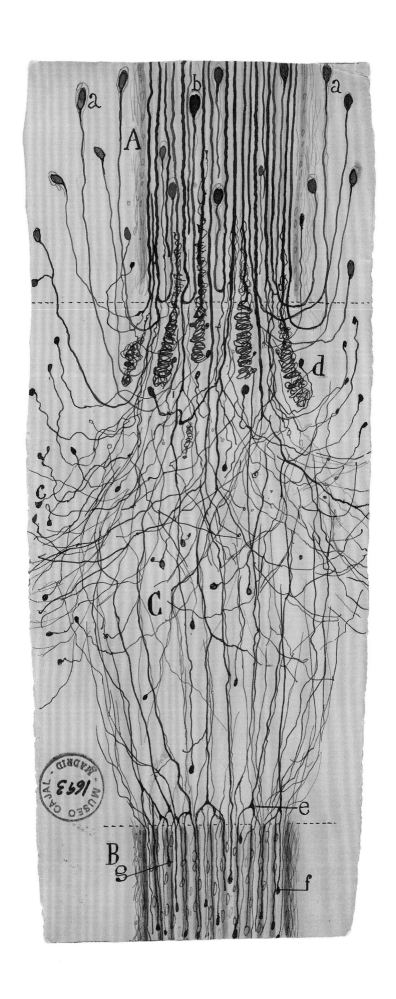

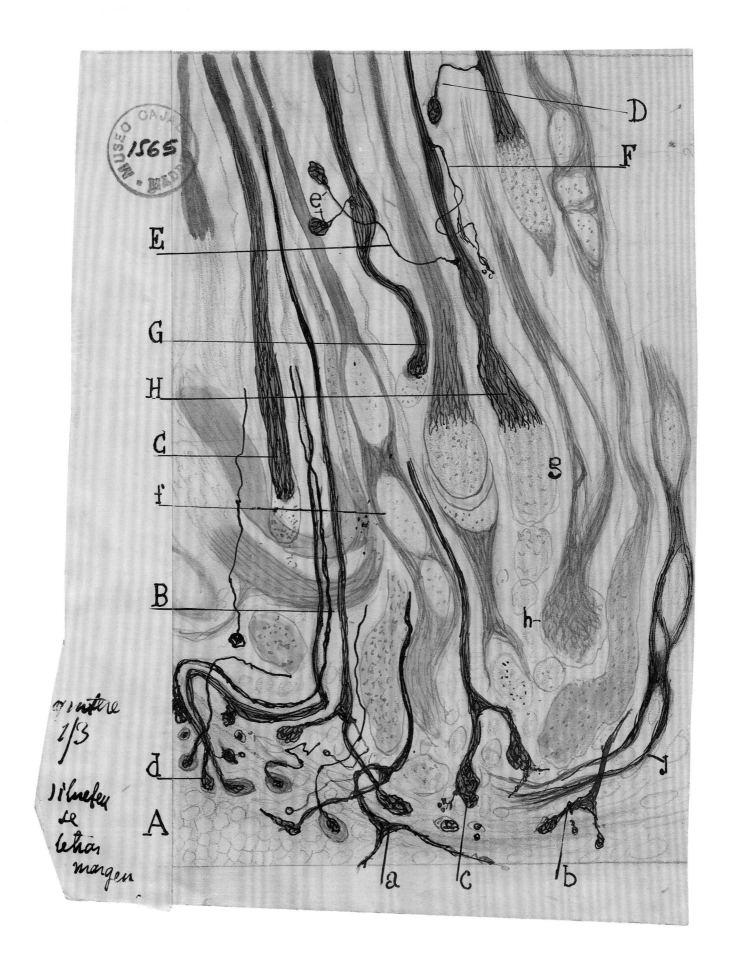

A cut nerve stump of the rabbit six hours after damage

Cajal documented the regrowth of peripheral nerves at different times after they were severed. He found signs of nerve regeneration as early as six hours after the nerves were cut. Here, label A indicates the initial wound, where the nerve was cut, while G and H show individual severed axons that are swollen due to the injury. Thin axon branches have sprouted from some of the axons (D, E, and F), the first indication of regeneration. Some of these axon branches (d) may grow across the gap separating the nerve stump and the disconnected end of the nerve (see page 179) and eventually make their way back to their target organ. If enough axons regrow through the peripheral nerve, nerve function will be restored.

Glial cells in a cut nerve separated from the spinal cord

When a nerve is cut, the axons in the disconnected portion of the nerve are separated from their cell bodies and, as a consequence, die away. However, the glial cells (called Schwann cells in peripheral nerves) that surround individual axons in the nerve and insulate them remain alive. This drawing illustrates these glial cells, which are colored darkly, in the separated portion of the nerve. Label f indicates a single glial cell. It ends at label b, a gap (the node of Ranvier) separating it from the next glial cell. Other glial cells in this drawing, including the one above f, have partially deteriorated into several pieces. As Cajal observed, when an axon regrows from the central stump of a cut nerve and makes its way to the separated portion of the nerve, it will regrow through these glial cells. Cajal's experiments on cut nerves revealed that the glial cells in the separated nerve release chemicals that encourage the axons to regrow. We now call these chemicals growth factors.

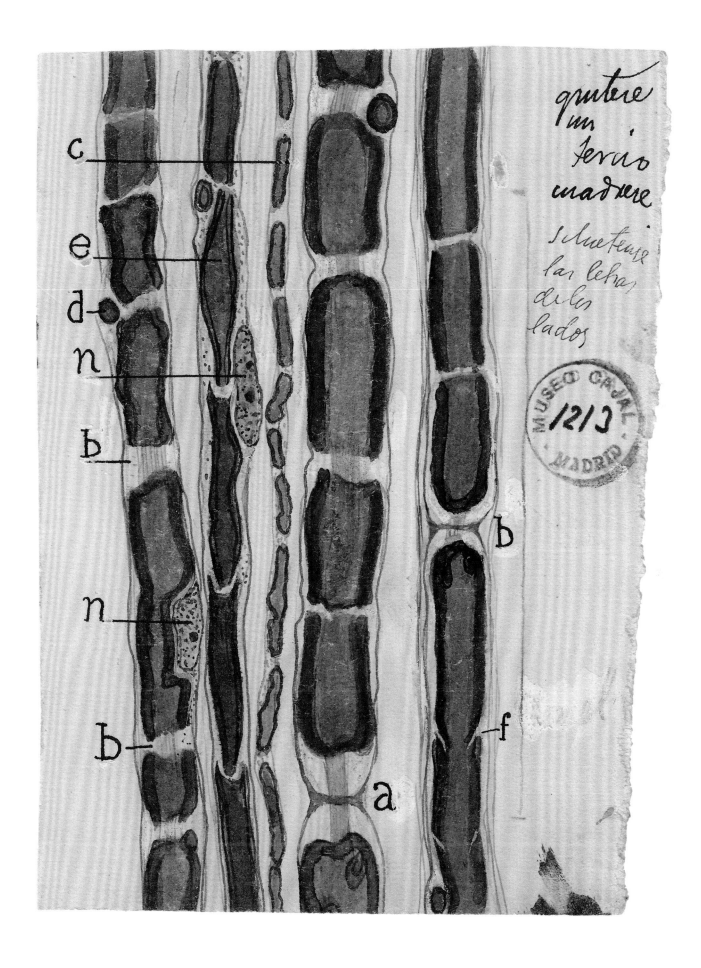

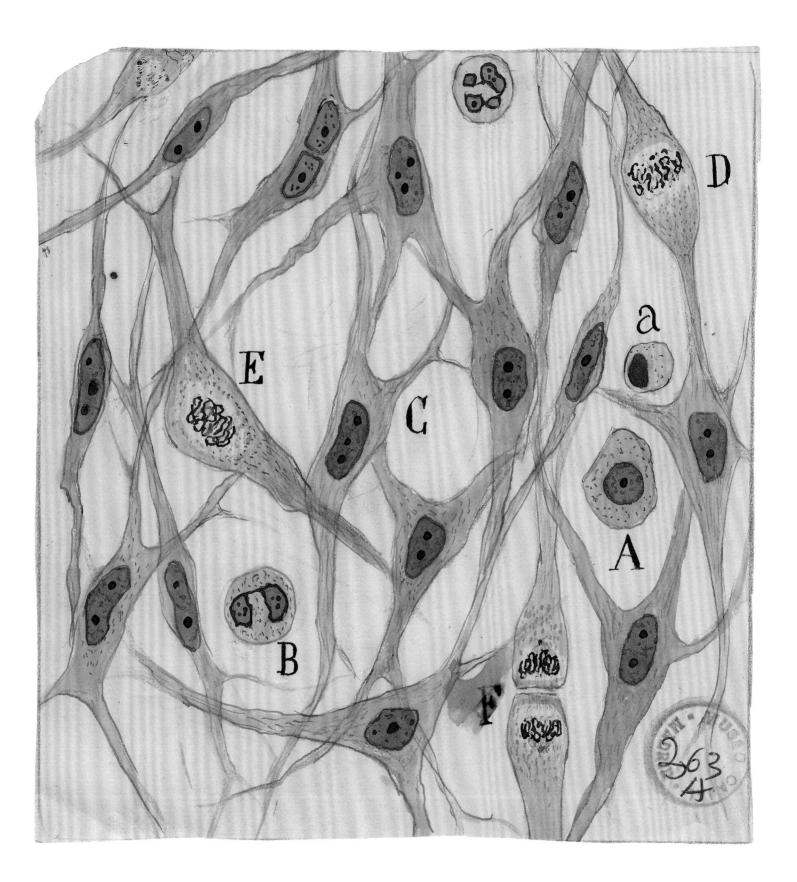

Scar tissue in a cut nerve stump

Scar tissue forms when a nerve outside the spinal cord is cut. This scar tissue is largely composed of fibroblasts, cells that play a critical role in wound healing. Cajal shows these scar tissue fibroblasts in this drawing of the central stump of a cut nerve still connected to the spinal cord. These cells divide rapidly, and Cajal illustrates the cells in different stages of division (D, E, and F). In his experiments on cut nerves, Cajal demonstrated that axons in the central stump were able to grow through the spaces between the fibroblasts in the scar tissue and reconnect to the disconnected portion of the nerve. This process allows nerves outside the spinal cord to regenerate and to regain function.

Axons in a damaged peripheral nerve entering the spinal cord of the cat

After a peripheral nerve is damaged, it is able to regenerate and reestablish normal function. The axons will regrow past the damaged portion of the nerve and reconnect to the skin or muscle. The axons can also grow back into the spinal cord. This drawing shows axons from a peripheral nerve four days after it was damaged. The damaged nerve crosses diagonally from the lower left, and the spinal cord is on the right. Cajal has shown that some axons within the nerve (a, c, and C) have regrown back into the spinal cord.

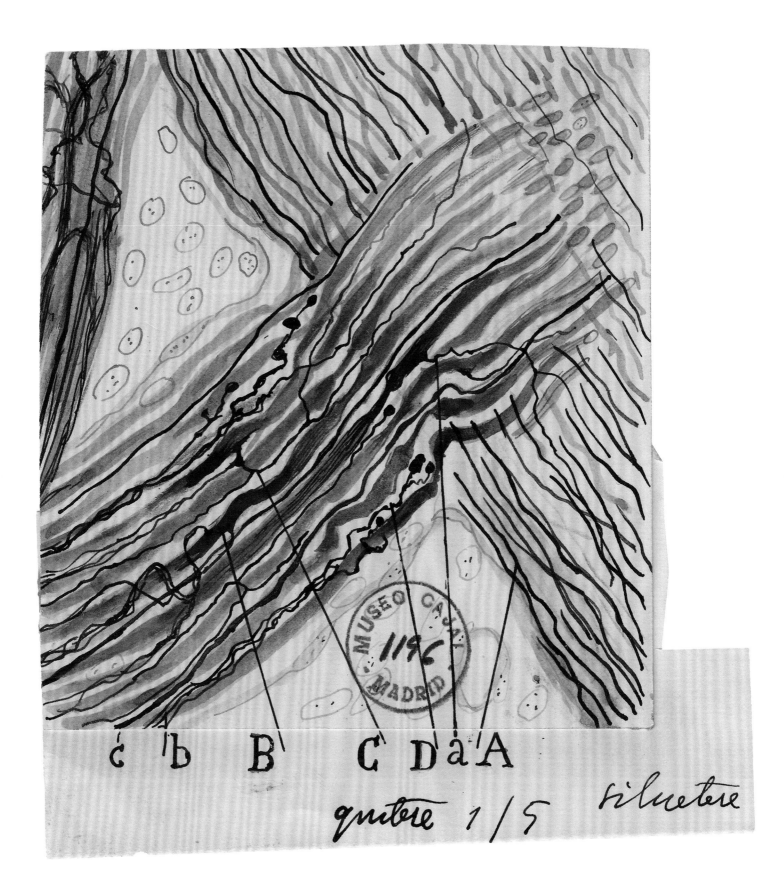

quizás
1/4
silueta

White blood cell migration across a blood vessel

When the brain or another organ is injured or infected, white blood cells migrate from within blood vessels (shown on the left) into the injured tissue (shown on the right) to fight infection and promote repair. Cajal illustrates this process here. In a sequence of seven steps, Cajal shows a white blood cell squeezing its way through a pore (A) in the blood vessel wall (B). The white blood cell can assume amoeboid properties, enabling it to crawl across the wall of the vessel (d, e, and f) and through the tissue (g). Although white blood cells infiltrate the brain when it is injured, the nervous system also possesses intrinsic cells that serve functions similar to those of white blood cells. These cells, called microglial cells, help fight infection and repair the brain following injury.

Tumor cells of the covering membranes of the brain

The surface of the brain is covered with a protective lining called the meninges (composed of three membranes: the dura mater, the arachnoid, and the pia mater). In a rare condition, a tumor of the meninges, called a dural endothelioma, can develop. These tumors are usually benign and often have no symptoms. In this drawing, Cajal illustrates dural endothelioma cells, which have a whorl-like appearance typical of this type of tumor. It is not clear whether Cajal was aware of the paintings of Vincent van Gogh, but this drawing strongly resembles van Gogh's depictions of the night sky in paintings such as *The Starry Night*.

SEEING THE BEAUTIFUL BRAIN TODAY

Janet M. Dubinsky

Santiago Ramón y Cajal used drawing both to illustrate his observations and to convey scientific arguments. The beauty of his plates helped convince other European neuroanatomists of the veracity of his conclusions, and it is clear from Cajal's writings that he understood the persuasive power of images. A stunning image is not easily forgotten—nor are the principles that it illustrates. The composition and clarity of Cajal's presentations added an aesthetic sensibility and a subtle, emotional appeal to their scientific content. Today, neuroscientists with access to vastly more complex visualization tools than Cajal had, running laboratories with teams of scientists and substantially more resources, also need to create images (digital, not hand drawn) to share observations and make arguments. Scientists today also deploy visual strategies to invest their images with emotional appeal. This essay presents images made in the last several years that illustrate what contemporary neuroscientists can show us about the brain.

Cajal focused on the cellular structures that comprise the nervous system. He worked at the microscopic scale, resolving parts of cells as small as 1 micron (μm, or one thousandth of a millimeter) and whole cells of 100 μm (the diameter of a medium-thick human hair) or more. Today, neuroscientists continue to work at the cellular scale explored by Cajal (opposite and page 195), but we also observe the brain at scales that were not accessible to him. Thanks to electron microscopy, we can probe much smaller scales, examining the structures within and interactions among parts of cells, synapses, and individual protein molecules (0.1 to 0.0001 μm, pages 197 and 198). On a much larger scale, advanced magnetic resonance imaging (MRI) techniques permit noninvasive imaging of the whole, living human brain (~0.1 to 20 cm, pages 199 and 200).

Cortex of a mouse genetically engineered to express randomly generated colors in every neuron

Neurons and all their dendrites, axons, and connections form an intertwined set of networks for information processing in the brain. Even when they are colorized, individual cells are difficult to follow through this crowded space. In the tightly packed areas where axons and dendrites interact, the almost impressionistic mix of colors ironically becomes grayish. The cortex, called gray matter, appears darker in fresh tissue than the bundles of axons that connect different parts of the brain, which are termed white matter.

At each scale, the scientific questions being asked probe different aspects of brain function. At the cellular scale, we investigate the interconnectedness of small groups of neurons, exploring the way incoming information is filtered and combined by a neuron to represent information at the next, higher processing level. At the synaptic and molecular level, we look at how neurons transform and transmit electrical and chemical signals at synapses—what Cajal called "intimate connections." And at the whole brain level, we regularly use MRI clinically, to probe the structural integrity of the brain in health and disease. Moreover, we begin to look at how the brain produces thoughts and behaviors, by identifying the neural networks that generate specific functions.

One thing that Cajal could not do was observe living brains in action, unlike his contemporaries, Sir Charles Scott Sherrington and William James, who pioneered the fields of neurophysiology and psychology, respectively. However, Cajal examined development and growth of the brain as well as degeneration and the limited regeneration that occurs after injury. All of these perspectives informed his projections of how the nervous system functioned. Today, we are constantly inventing new techniques at all scales to observe and quantify the brain's electrical and chemical activity while subjects, both humans and animals, are performing mental and physical tasks. Contemporary neuroscience concentrates on understanding how the structures described by Cajal produce the myriad functions and behaviors they support.

CELLULAR SCALE

In the late nineteenth and early twentieth centuries, Cajal adapted and created techniques in microscopy and histology to visualize the fine structure of cells in the nervous system. Today's visualization tools and techniques exploit the biological, pharmacological, computational, and engineering achievements of the twentieth century. For example, we can use genetic engineering to place proteins in specific cells, which causes those cells to fluoresce in different colors when light strikes them. One image that expresses both the beauty and complexity of the nervous system uses this technique to illuminate the neurons in the brain—this is the so-called Brainbow mouse (see page 192).[1] Here, the neurons in the cortex are fluorescing a rainbow of approximately one hundred colors that highlight the densely packed cortical environment. Ideally, this technique could be used to trace a single neuron through the forest of other brain neurons, but this has proved to be challenging to achieve with light microscopy.

More targeted techniques using fewer colors are routinely applied to identify neurons in specific circuits. In one experiment, a cleverly designed virus injected into a single cortical neuron in a mouse's brain colored that neuron pink and all the neurons that send information to it green (opposite).[2] While Cajal could infer which classes of neurons talked to one another (see, for example, page 123), he could not identify the complete set of individual neurons that connected to any one specific cell. Fluorescent proteins that change

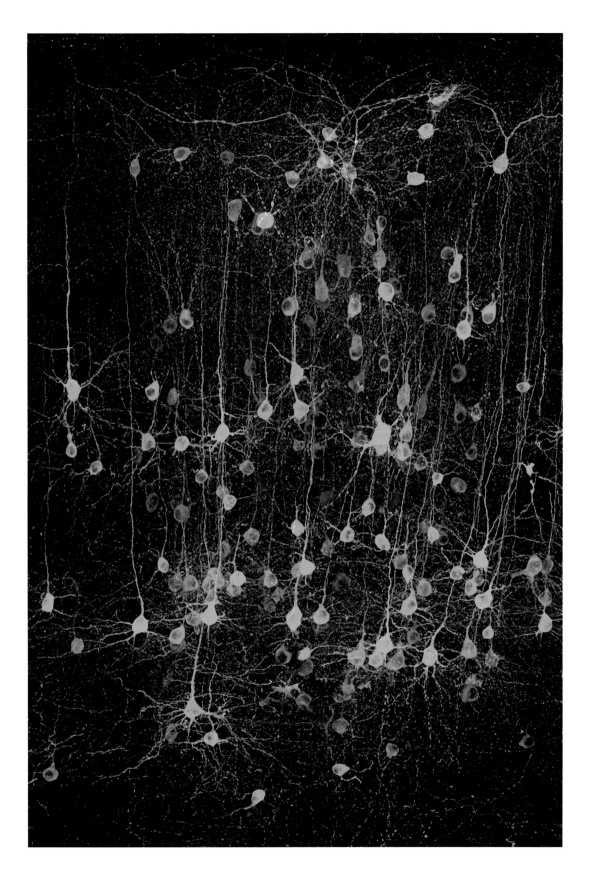

Convergence of inputs onto a single neuron

In this section of mouse cortex, the pink neuron receives input from each and every green neuron. Most of the processes on all of these neurons are dendrites, the input ends of neurons. The slightly thicker points along the pink dendrites may represent spines, sites of incoming synapses. The pink axon descends among the green cells, eventually leaving the image. Brighter spots along the axon may be synapses onto neurons that do not contain colored proteins. Axons of the green neurons are harder to locate among the many dendrites. The density of green neurons is less than the density for all neurons this region, shown on page 192, indicating that not all neurons connect to the pink neuron.

SEEING THE BEAUTIFUL BRAIN TODAY

their brightness during neuronal activity have also been developed. Using markers like these, we can measure the activity of populations of neurons while the animal subject engages in a precise behavior under the microscope.

In the experiment illustrated on page 195, all of the neurons pictured were also made to produce a third colored protein (not visible in this image) whose intensity varied in time with the cells' electrical and chemical activity. The mouse viewed stripes moving across a screen, while the flow of information from the green neurons into the pink receiving neuron was recorded. An experiment like this reveals how information is combined and filtered from one set of neurons to the next, a process that is critically important for unraveling how the brain creates the mind. Cajal was able to draw the complexity of cortical connections, but he lacked the means to view and actually measure electrical activity in living brains over time.

SYNAPTIC AND MOLECULAR SCALES

Connections between neurons constitute over 90 percent of the volume of the cortex. One of contemporary neuroscience's biggest challenges is to try to find patterns or rules within the apparent mad scramble of tightly packed cells and their synapses. This quest is behind the current push to define the connectome—that is, to map the connections among all the neurons in a brain. The goals of today's connectome projects range from detailed investigations of the synapses where neurons talk to one another to using big data approaches to produce new insights about how the brain as a whole functions. The resulting knowledge might then be exploited in treatments for mental illness and diseases of the nervous system.

One class of connectome experiments seeks to reconstruct all the synapses within a small piece of mouse cortex (opposite). The tissue is sliced extremely thinly (0.03 μm thick) in an automated process that sequentially captures and images each slice with an electron microscope. Using electrons instead of light increases the resolution of the images almost to the level needed to view individual proteins, capturing internal neuronal structures and the spaces between neurons that Cajal could not see. Each successive image is colorized, so that when all the images are stacked sequentially, an individual axon or dendrite can be followed down through the stack as it twists and turns around its neighbors. Computers are programmed to use artificial intelligence strategies to look into the stack and reconstruct three-dimensional views of these structures. When the computer program can't decide which path to follow, scientists and citizen-scientists may become engaged to manually perform the tracing, forming a human-computer collaboration.[3] Multiple ironies abound in this process that uses humans to teach (that is, program) computers to make decisions based upon human-derived understanding of how brains operate, all to increase our understanding of the individual connections among neurons in brains.

Reconstructing tiny regions of the cortex using this method is arduous.[4] From such heroic efforts, we learn that not every axon synapses upon every dendrite that it crosses.

3-D view of synapses on several spines along a cortical dendrite

This digital image represents a reconstruction from serial section electron micrographs of a dendrite and its surrounding axons in a mouse cortex. The red dendrite traverses the image from lower right. The heights of the spines (solid red, center) are approximately 1μm. The synaptic vesicles (small white spheres) that release chemical messages at each synapse are shown in the axon branches (transparent colors) surrounding the dendrite. Synapses strengthen or weaken with practice or disuse, a property that underscores learning at the cellular level. This variability is referred to as synaptic plasticity, an idea Cajal embraced as necessary for mental function. The need for such plasticity was one of Cajal's main arguments against the Reticular Theory. The authors of this image used a bright, reflective color scheme, which gives the surfaces a plastic-like texture. Consider how plastic, like synapses, can be molded into an infinite number of shapes.

Connections between neurons in the cortex appear to be targeted and specific in ways we still do not understand. To probe this question further, we will need to examine the variability across multiple cells and between different individuals. The task is daunting. Considering the time and effort expended by teams of investigators on these massive undertakings, the synthesis Cajal achieved with his daily rhythm of viewing, contemplating, and drawing remains a remarkable achievement.

SEEING THE BEAUTIFUL BRAIN TODAY

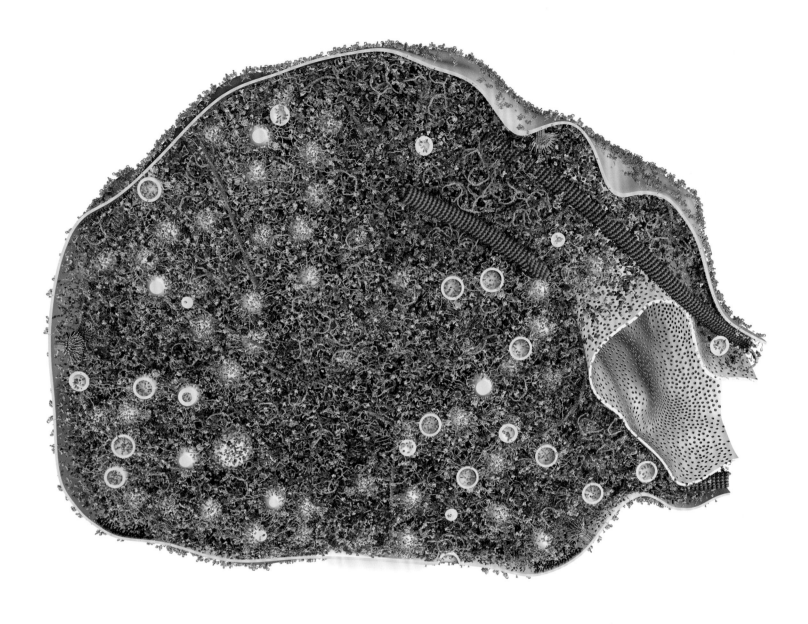

Longitudinal view through a nerve terminal

This digital reconstruction of the very end of an axon portrays the shapes and sizes of the six hundred most common proteins found in the nerve terminal in their correct proportions and positions.[3] The red line at the bottom of the image marks the location of the synapse. When an electrical pulse arrives at the nerve ending, chemical messengers packaged inside the white vesicles are released at the synapse, an intricate dance coordinated by many of these proteins working together. Additional proteins near the small indentation to the right of the synapse retrieve, recycle, and refill the vesicles. The nerve ending is approximately 0.6 μm high by 1 μm wide; the axon would extend off to the right. The white polka-dotted hollow membrane in the fold to the right is a mitochondrion whose inner structure was not included in this model.

A central tenet of contemporary neuroscience is the flexibility of information transfer across synapses. Cajal established the existence of these transfer points or "intimate connections," although he could not resolve the space between an axon and a dendrite. Still, these points of contact between neurons implied to him that a mechanism must exist for communication across this space, and he hypothesized correctly that the synapses would turn out to be the most variable part of the nervous system, changing dynamically to support the growth, learning, and experiences that fill a life. Understanding the molecular machinery that translates the electrical signals of neurons into the chemical messages that cross synapses has been a primary challenge of the past quarter century. Numerous techniques have been applied to explore nerve endings on a molecular level. Chemists use genetic and crystallographic information to construct models of individual proteins. Further measurements determine how many copies of each protein are found in nerve endings. Highly detailed images of nerve endings made with electron microscopes reveal the relationships among structures within nerve endings. All of these have now been combined into a molecular-level view of a nerve ending where the chemical messages are released (opposite).[5]

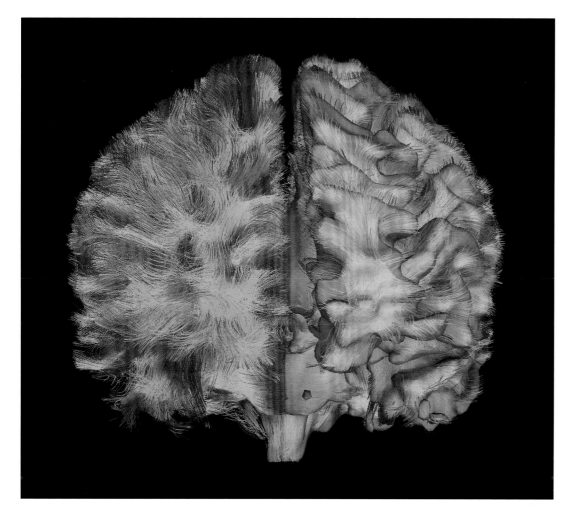

Nerve fibers in the white matter form connections between different parts of the brain, viewed from the front

Each line represents a bundle of axons running in parallel. These bundles are called nerve fibers. The nerve fibers that make up the brain's white matter are shown on the left. On the right, a folded layer representing the boundary where the gray matter ends and the white matter begins has been superimposed. The fibers extending above this surface represent axons entering or leaving the cortical gray matter. Colors represent the direction that the fibers travel: Green indicates front (nose) to back of head; red, left to right; blue, top of the head to bottom (below chin). Other colors indicate fibers traveling at angles, or combinations of those principal directions.

SEEING THE BEAUTIFUL BRAIN TODAY

This image, like Cajal's drawings, provides a static view of a dynamic structure. It represents one of the transparent axonal structures filled with vesicles that appear as tiny white dots or spheres in the image on page 197. Proteins float in the watery environment inside these nerve endings. Some are tethered together, ready for coordinated action when an electrical pulse arrives. Then, the vesicles, which contain chemical messages, are guided and manipulated by some of these proteins toward the synapse, the bright red area at the bottom of the image. At the synapse, the vesicles fuse with the surface membrane and dump their contents into the restricted space between neurons. The chemicals cross this space (~0.1 μm) and bind to proteins on the dendritic membrane (not shown), initiating an electrical signal in the receiving neuron. This chemical communication across a synapse takes several thousandths of a second, and the recycling of the vesicles can be equally rapid or may proceed for over a minute or more. Millions of synapses contribute to simple intentional behaviors. As we practice the behavior, the protein machinery at synapses becomes more efficient, and the synaptic signals become stronger. New synapses form and unused ones retract. These subtle molecular and structural changes provide the plasticity required for learning and memory.

WHOLE BRAIN SCALE

Ultimately, questions about how the brain produces the mind must be answered with studies of the human brain performed on subjects engaged in measurable behaviors in controlled settings. Advanced MRI techniques, including diffusion tensor imaging (DTI), are used for noninvasive imaging of both the normal and diseased human brain. Structural MRI images precisely resolve the gray matter (areas with cell bodies and dendrites) and the white matter (areas of axons or nerve fibers) visible in the gray slices in the image on page 201. Further processing can reveal nerve fibers, or bundles of axons, interconnecting distant parts of the brain in three dimensions, as in the image on page 199. DTI "sees" these long fiber bundles by mapping the movement of water molecules inside axons. Visualizing these fibers emphasizes the constantly active information highways that connect different parts of the brain.

To identify the brain regions involved in complex behaviors, scientists combine functional measurements with the structural brain images (opposite).[6] The subject performs some cognitive task, say, game playing, by looking at a screen and pressing a button. During the game, brain areas with more neural activity require more energy, which is delivered by increasing blood flow. Functional MRI (fMRI) is used to measure blood flow, indicating brain regions involved in generating the thoughts or behaviors. Regions of increased blood flow are represented by colored areas superimposed on structural images. These functional images can now be generated in less than a second, covering the entire brain. This is not nearly as fast as the electrical signals in neurons (one thousandth of a second), but fast enough to examine regional brain activity as plans and decisions are being made and considered.

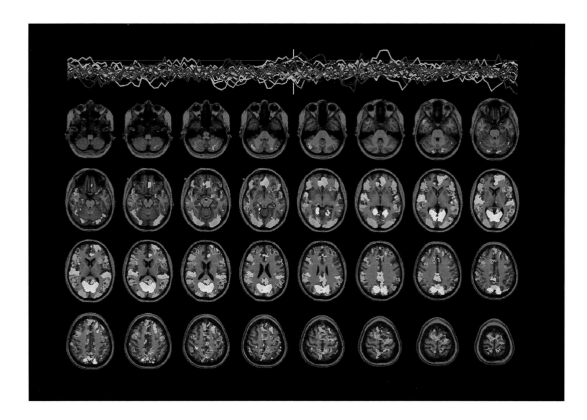

Many regions of the brain are simultaneously active, even during rest

This set of rapidly gathered fMRI images depicts brain regions active during normal awake but resting behavior. Subjects could be daydreaming or thinking about anything they chose, but were not directed to perform any specific mental tasks by the experimenters. The fMRI activity of a set of brain regions monitored over time is shown in the traces at the top. Structural images of horizontal slices of the subject's brain are ordered from bottom of the brain (top row, left) to top of the head (bottom row, right). Colored patches on the images represent regions of functional activity outside the average range at the time marked by the white line. Gray areas represent brain regions with average activity levels.

The work of many laboratories has generated data that points to consistent patterns of activation among brain regions when subjects are executing tasks such as moving a limb, adding numbers, hearing, reasoning, and responding emotionally. We now know that single regions of the brain are rarely activated alone. Instead, specific behaviors arise from the coordinated activation of sets of regions. These sets are classified as networks. Networks appear to be active during behaviors and also during rest. Pathways of axon bundles crossing between brain regions, so meticulously traced by Cajal (see pages 111, 128, and 142) and visualized with DTI (see page 199), provide the anatomical infrastructure for the constantly changing networks of activity detected with fMRI (above). Computer analysis of time-varying fMRI has become refined enough so that some patterns of neural activation can actually tell us in broad terms what the subject is thinking about.

In the century since Cajal identified neurons as individual units carrying signals through cellular circuits, neuroscience has progressed. We can "see" more of the brain than Cajal was able to with the tools he possessed. We now understand in fine, but imperfect, molecular detail how the neurons and synapses he identified send signals across biological space. But we still do not understand the connectome, the brain's full "wiring diagram," which may vary by individual experiences. We do not know how the brain, three pounds of water, oil, small molecules, and protein, can perform so many individual computations using so little energy. We do not fully understand how the brain creates the mind. In many ways, our twenty-first-century scientific questions and goals remain the same as Cajal's: to "clarify the secret of mental life."[7]

NOTES

"SANTIAGO RAMÓN Y CAJAL"
by Larry W. Swanson (pages 11–18)

[1] Santiago Ramón y Cajal, *Recollections of My Life*, translated by E. Horne Craigie with Juan Cano (Cambridge: MIT Press, 1989), 291.

[2] Ibid., 321.

[3] Ibid., 325.

[4] See G.M. Shepherd, *Foundations of the Neuron Doctrine*, 25th anniversary edition (Oxford: Oxford University Press, 2016).

[5] For the bird cerebellum, see Santiago Ramón y Cajal, "Estructura de los centros nerviosos de las aves," *Revista trimestral de Histología normal y patológica* 1 (1888): 1–10. For Cajal's systematic confirmation of his principles, see Santiago Ramón y Cajal, *Histologie du système nerveux de l'homme et des vertébrés*, translated by L. Azoulay (Paris: Maloine, 1909, 1911). For an English translation, see Santiago Ramón y Cajal, *Histology of the nervous system of man and vertebrates*, translated by Neely Swanson and Larry W. Swanson (New York: Oxford University Press, 1995).

"DRAWING THE BEAUTIFUL BRAIN"
by Lyndel King and Eric Himmel (pages 21–28)

[1] Cajal told a journalist in 1900 that he had made more than 12,000 drawings, but about 2,900 remain.

[2] Cajal, *Recollections*, 17.

[3] Ibid., 36.

[4] Ibid., 169–170.

[5] Ibid., 83, 146–147, 278.

[6] As Javier DeFelipe, the leading expert in the anatomical illustrations of these great nineteenth-century neuroscientists, writes, "The observer is required, in the act of drawing, to highlight details he or she considers important: A key feature for one scientist can pass unnoticed by another, and two expert anatomists drawing from the same sample could potentially produce radically different diagrams." See Carl Schoonover, "The Birth of Modern Neuroscience: Santiago Ramón y Cajal," *Portraits of the Mind* (New York: Abrams, 2010), 52.

[7] See Gunnar Grant, "How Golgi Shared the 1906 Nobel Prize in Physiology or Medicine with Cajal," Nobelprize.org, September 12, 1999, http://www.nobelprize.org/nobel_prizes/medicine/laureates/1906/article.html.

[8] Some authors hold that Cajal made virtually all of his illustrations from memory, rigidly segregating observation time from drawing time, but there is also testimony from colleagues and students who watched him drawing at the microscope. For a fascinating look at degrees of variance from Cajal's histological slides to his drawings, see Pablo Garcia-Lopez, Virginia Garcia-Marin, and Miguel Freire, "The Histological Slides and Drawings of Cajal," *Frontiers in Neuroanatomy* 4 (March, 2010): Article 9.

[9] Milton Glaser and Judith Thurman, *Drawing Is Thinking* (New York: Overlook Press, 2008).

[10] Al Tauber, ed., *The Elusive Synthesis: Aesthetics and Science* (New York: Springer, 2012), 60. Cajal is quoted in Laura Otis, *Membranes* (Baltimore: Johns Hopkins University Press, 2000), 83–84.

[11] Otis, *Membranes*, 84.

[12] Cajal, *Recollections*, 338.

[13] Kelly Minner, "Interview with Frank Gehry," April 21, 2011, *ArchDaily*, http://www.archdaily.com/129680/interview-with-frank-gehry. For a discussion of Gehry's drawings and *disegno*, see Horst Bredekamp's introduction to Mark Rappolt et al., *Gehry Draws* (Boston: MIT Press, 2004).

[14] Charles Scott Sherrington, "A Memoir of Dr. Cajal," in *Explorer of the Human Brain: the Life of Santiago Ramón y Cajal*, Dorothy F. Cannon (New York: Henry Schuman, 1949), xiii–xiv.

[15] See Schoonover, *Portraits of the Mind*, 59.

[16] See Santiago Ramón y Cajal [writing as Dr. Bacteria], "The Corrected Pessimist," in *Vacation Stories*, translated by Laura Otis (Champaign, Illinois: University of Illinois Press, 2001).

[17] An exhibition exploring the Surrealists' interest in Cajal's drawings took place at the University of Zaragoza, Spain, in 2015: *Fisiología de los sueños. Cajal, Tanguy, Lorca, Dalí*.

[18] Santiago Ramón y Cajal, *El mundo visto a los ochenta años: impressiones de un arteriosclerótico* (Madrid: Artistica, 1934).

COMMENTARIES ON THE DRAWINGS
(pages 34–191)

[1] Santiago Ramón y Cajal, "Estructura intima de los centros nerviosos," *Rev. Ciences Med.* 20 (1894): 159–160.

[2] Cajal, *Recollections*, 364.

[3] Santiago Ramón y Cajal, "Contribución al conocimiento de la neuroglia del cerebro humano," *Trabajos del Laboratorio de Investigaciones Biológicas, Tomo XI* (1913), 255–314, 313.

[4] Ibid., 261.

[5] Santiago Ramón y Cajal, "Algunas conjeturas sobre el mecanismo anatómico de la ideación, asociación y atención," *Revista de Medicina y Cirugía Prácticas* (1895), 11.

[6] Cajal, *Recollections*, 576.

[7] Ibid., 383–384.

[8] Ibid., 472.

[9] Santiago Ramón y Cajal, "Neuron Theory or Reticular Theory," translated by M. Ubeda Purkiss and Clement A. Fox (Madrid: Instituto Ramón y Cajal, 1954), 29.

[10] Cajal, *Recollections*, 389.

[11] Sherrington, "A Memoir of Dr. Cajal," xii.

[12] Cajal, *Recollections*, 363.

[13] Ibid., 415.

[14] Ibid., 414.

[15] Sherrington, "A Memoir of Dr. Cajal," xiii.

[16] Cajal, *Recollections*, 369.

"SEEING THE BEAUTIFUL BRAIN TODAY"
by Janet M. Dubinsky (pages 193–201)

[1] Jean Livet, Tamily A. Weissman, Hyuno Kang, Ryan W. Draft, Ju Lu, Robyn A. Bennis, Joshua R. Sanes, and Jeff W. Lichtman, "Transgenic strategies for combinatorial expression of fluorescent proteins in the nervous system," *Nature* 450 (2007): 56–62.

[2] A. Wertz, S. Trenholm, K. Yonehara, D. Hillier, Z. Raics, M. Leinweber, G. Szalay, A. Ghanem, G. Keller, B. Rozsa, K.K. Conzelmann, and B. Roska, "Presynaptic Networks. Single-cell-initiated monosynaptic tracing reveals layer-specific cortical network modules," *Science* 349 (2015): 70–74.

[3] See www.eyewire.org.

[4] Narayanan Kasthuri, Kenneth J. Hayworth, Daniel R. Berger, Richard L. Schalek, José A. Conchello, Seymour Knowles-Barley, Dongil Lee, Amelio Vázquez-Reina, Verena Kaynig, Thouis R. Jones, Mike Roberts, Josh L. Morgan, Juan C. Tapia, H. Sebastian Seung, William G. Roncal, Joshua T. Vogelstein, Randal Burns, Daniel L. Sussman, Carey E. Priebe, Hanspeter Pfister, and Jeff W. Lichtman, "Saturated Reconstruction of a Volume of Neocortex," *Cell* 162 (2015): 648–661.

[5] B.G. Wilhelm, S. Mandad, S. Truckenbrodt, K. Krohnert, C. Schafer, B. Rammner, S.J. Koo, G.A. Classen, M. Krauss, V. Haucke, H. Urlaub, and S.O. Rizzoli, "Composition of isolated synaptic boutons reveals the amounts of vesicle trafficking proteins," *Science* 344 (2014): 1023–1028.

[6] Stephen M. Smith, Karla L. Miller, Steen Moeller, Junqian Xu, Edward J. Auerbach, Mark W. Woolrich, Christian F. Beckmann, Mark Jenkinson, Jesper Andersson, Matthew F. Glasser, David C. Van Essen, David A. Feinberg, Essa S. Yacoub, and Kamil Ugurbil, "Temporally-independent functional modes of spontaneous brain activity," *PNAS* 109 (2012): 3131–3136.

[7] Cajal, *Recollections*, 363.

INDEX

A

Advice for a Young Investigator (Cajal), 18
anatomy
    examination of, 13, 23
    in art history, 21
Aristotle, 14
artistic talent, 24, 26–28, 202
astrocytes
    after injury, 174, *175*, *176*, 177
    in cerebral cortex, 64, 65, *68*, 69
    in hippocampus, 66, *67*, 70, *71*, *72*, 73
    in retina, 88, *89*
    in spinal cord, 74, *75*, *76*, 77
auditory system
    in brain, *104*, 105, 106, *107*
    in inner ear, *100*, 101, 102, *103*
autobiography, 12–18
axons
    in injured peripheral nerve, 178, *179*, *180*, 181, 186, *187*, *188*, 189
    in spinal cord, *148*, 149, 150, *151*
    of injured Purkinje neurons, 164, *165*
    of pyramidal neurons, 42, *43*

B

birth, 12
blood vessels
    glial cell association with, 66, *67*, 74, *75*, 174, *175*, *176*, 177
    white blood cell migration across, *188*, 189
brain
    composition of, 35
    early examination of, 8–9, 14, 23, 34
    information flow in, 84–85, *86*, 87, 122–23, *130*, 131, 132, *133*, *134*, 135, 136, *137*, 140, *141*, 194
    MRI, *200*, 201
    organization of, 9, 11, 35
    structures within, 35
    theories, composition of, 146, *147*
    tumor cells in, 190, *191*
brain activity, *201*
brainstem
    medulla, 144, *145*
    neurons of, *104*, 105, *108*, 109

C

Cajal, Santiago Ramón y. *See* specific topics
calyces of Held
    in nucleus of trapezoid body, 31, *104*, 105
caudate nucleus
    in basal ganglia, *128*, 129
cerebellum
    after injury, 170, 171
    of cat, photomicrograph, *120*
    Purkinje neurons of, 48, 49, 50, *51*
    stellate neurons in, *52*, 53, 54, *55*
cerebral cortex
    astrocytes of, 64, 65, 68, 69
    auditory cortex in, 106, *107*
    dendrites of pyramidal neurons, 44, 45
    glial cells of, 64, 65, 68, 69, 174, *175*, *176*, 177
    injury to, 172, *173*, 174, *175*, *176*, 177
    of genetically engineered mouse, *192*, 193, *195*, *197*
    pericellular nests in, 46, *47*
    pyramidal neurons of, *36*, 37–38, *39*, 40, 41–42, *43*, 44, 45
Charlas de café (Cajal), 18
children, 15
cuneate nucleus
    neurons within, *152*, 153

D

Darwin, Charles, 11
da Vinci, Leonardo, 21
death, 18
De corporis humani fabrica libri septem (Vesalius), 13
dendrites, 9, 35
    of pyramidal neurons, 44, 45
    of Purkinje neurons, 48, 49, 50, *51*
    of retinal ganglion cells, 96–97
development
    stages of, neuron, 159
diffusion tensor imaging (DTI), 200
dorsal root ganglia, 118, *119*, 132, *133*

E

ear. *See* inner ear
early life, 12–13
early technique
    for examination of brain, 8–9, 14, 23, 34, 45
Ehrlich, Paul, 45
El mundo visto a los ochenta años (Cajal), 18
Estrategia lapidaria (Cajal), 13
Estudios sobre la degeneración y regeneración del sistema nervioso (Cajal) (1913, 1914), 18
examination
    current technology for, 193–201
    early technique for brain, 8–9, 14, 23, 34, 45
    of anatomy, 13, 23
eyes, 84–97

F

family and friends (1885), 6–7
fly
    cells from optic lobe of, 78, *79*
friend, in late teens (1870), *20*, 21

G

Galvani, Luigi, 9
Gehry, Frank, *26*
glial cells
    in cerebral cortex, 64, 65, *68*, 69, 174, *175*, *176*, 177
    in hippocampus, 66, 67, 70, *71*, *72*, 73
    in peripheral nerve, 182, *183*
    in spinal cord, 74, *75*, *76*, 77
Golgi, Camillo, 8–11
    stain method by, 14–16, 45
growth cones, 159, 160, *161*
gut
    Cajal neurons in, *60*, 61

H

hippocampus
    astrocytes in, 66, *67*, 70, *71*, *72*, 73
    connections within, 136, *137*, 140, *141*
    pyramidal neurons in, 136, *137*, *138*, 139, 140, *141*
    structure and connections of, 136, *137*, *138*, 139, *142*, 143
histology, 13–14

I

information flow, 18
    from eyes, *86*, 87
    in brain, 84–85, *86*, 87, 122–23, 130–137, 140, *141*, 194
    in retina, 85, *94*, 95
injury
    astrocyte response to, 174, *175*, *176*, 177
    axon response to, 164, 172, *173*, 178, *179*, *180*, 181, 186, *187*
    cerebral cortex response to, 174, *175*, *176*, 177
    Purkinje neuron response to, 164, *165*, 166, 167, 168, *169*, 170, 171

pyramidal neuron response to, 172, *173*
inner ear, *100*, 101, 102, *103*

L
Lacabra, Luis Simarro, 14
lateral geniculate nucleus, 114, *115*
landscape with chapel (1871), *22*
Lorca, Federico García, 28, 29

M
magnetic resonance imaging, 193–95, 200–01
medial geniculate nucleus, *116*, 117
medulla, 144, *145*
microanatomy. See anatomy
midbrain
   neurons in, 98, *99*, 154, *155*
mouse
   cerebral cortex of, genetically engineered, *192*, 193, *195*, *197*
MRI. See magnetic resonance imaging
muscle cells
   in leg of scarab beetle, *80*, 81
   pathways for control of, *130*, 131, *134*, 135

N
nerve, peripheral
   regeneration of, *156*, 178, *179*, *180*, 181, *182*, *183*, *184*, 185, *186*, *187*
nerve fibers in brain, 199
nervous system, 16–18
Neuron Doctrine, 9, 18, 34, 53, 105, 122, 146, 158, 160
neurons, 35
   See pyramidal neurons, Purkinje neurons, stellate neurons
neuroscience
   father of modern, 8–9, 11
   nucleus of trapezoid body, 31, *104*, 105

O
olfactory bulb, *112*, 113
olfactory pathways, 110, *111*
optic lobe of the fly, 78, *79*

P
Pasteur, Louis, 11
pericellular nests
   around pyramidal neurons in cerebral cortex, 46, *47*
photoreceptor cells of retina, *90*, 91, *92*, *93*
portrait
   of wife, *15*
   of young girl (1868), *22*
   with children, *15*
Purkinje neurons
   after injury, 164, *165*, *166*, 167, 168, *169*, *170*, 171
   of cerebellum, *48*, *49*, 50, 51, 53
pyramidal neurons
   after injury, 172, *173*
   of cerebral cortex, *32*, *36*, 37, 38, *39*, 40, 41, 42, *43*, 44, 45
   of hippocampus, 136, *137*, *138*, 139

R
Ramón y Cajal, Santiago. See Cajal, Santiago Ramón y
Recollections of My Life (Cajal), 12–18
reflexes
   for vomiting and coughing, *134*, 135
regeneration
   of nerve fibers, *156*, 178, *179*, *180*, 181, *186*, *187*
Reticular Theory, 16, 26, 34, 146, 197

retina
   information flow in, 85, *94*, 95
   photoreceptor cells of, *90*, 91, *92*, *93*
   structure of, 88–97

S
scarab beetle
   muscle cells of, *80*, 81
scar tissue
   in cerebral cortex, 174, *175*, *176*, 177
   in cut nerve stump, *184*, 185
school
   of art, 22
   of medicine, 13
self-portrait
   at sixty-eight (1920), *10*, 11
   at thirty-four (1886), *2*, 7
   in laboratory in thirties (c. 1885), 24, *25*
   in late fifties (c. 1910), *4–5*, 7
   in thirties in library, *12*
skin cell
   division of, *162*, 163
small intestine
   neurons in, 58, *59*, *60*, 61
spinal cord
   axon fiber tracts in, *148*, 149, 150, *151*
   axons from damaged nerve into, *186*, 187
   glial cells in, 74, *75*, *76*, 77
   sensory pathways in, 132, *133*
stellate neurons
   in cerebellum, *52*, 53, 54, *55*
   in cerebral cortex, *124*, 125
still life
   of fruit and flowers (1912), 18, *19*
   of microscope and laboratory chemicals, *8*
street performers, 16, *17*
structure
   of brain, 35, *142*, 143, 144, *145*
   of neurons, 35, *56*, 57

superior cervical ganglion
   neurons in, 62, *63*
superior colliculus, 98, *99*
synaptic contacts
   in calyces of Held, 31, *104*, 105
   in cerebellum, *52*, 53
   in cerebral cortex, 196–99

T
technology
   for brain examination, 193–201
Theory of Dynamic Polarization, 9, 18, 122
tumor cells
   of brain, 190, *191*

V
Vesalius, Andreas, 13
vestibular nerve
   in brainstem, *108*, 109
vomiting and coughing
   reflexes for, *134*, 135

W
white blood cell
   migration across blood vessel, *188*, 189
wife, *15*

Y
young women on beach, *16*

ACKNOWLEDGMENTS

Santiago Ramón y Cajal was one of the founders of modern neuroscience. While his contributions to current concepts of brain function and organization are widely referenced in scientific publications, his carefully executed drawings—a rare mix of artistic skill and scientific insight—are much less known. It is therefore a great pleasure to share a representative sample of his drawings, which are held by the Cajal Institute, with the Weisman Museum of the University of Minnesota. The traveling exhibition planned by the museum is a perfect occasion to show the scientific community, and hopefully the American public at large, the beautiful organization of the nervous system as first acknowledged by modern science. Together with the highly motivated organizers of this well-planned initiative, we at the Cajal Institute are confident that this sample of the Cajal Legacy will be taken also as an esthetic experience for those who approach it for the first time. Science and are often travel together.

Ignacio Torres Alemán
Director, Cajal Institute
Madrid

The Spanish National Research Council (CSIC) is the largest public institution dedicated to multidisciplinary research in Spain and the third largest in Europe, aiming to promote Spain's competitiveness and foster scientific, educational, and economic development.

CSIC research is driven by its 132 centers and institutes, which are spread across the country, employing over 12,000 people, including more than 3,000 staff researchers and an equal number of contracted PhDs and other researchers. CSIC also manages numerous strategic national and international research facilities and has strong ties with universities, technology centers, hospitals, and non-profit organizations. The CSIC is connected to the productive sector through R&D contracts and technology transfer activities and remains the reference institution for collaborative projects and joint publications with Latin-American Universities.

After Cajal was awarded the Moscow Prize in 1900, his prestige led to the creation of the Laboratorio de Investigaciones Biológicas (LIB) the following year by order of HM the King Alfonso XIII. This research center was intended to provide the necessary help of the nation to Cajal studies. Following the award of the Nobel Prize in Physiology and Medicine in 1906, Cajal was appointed president of the Junta para Ampliación de Estudios e Investigaciones Científicas (JAE, 1907–1939) of the Spanish Ministry of Public Instruction and Fine Arts. Cajal encouraged critical changes in Spain's educational structures with the goal of modernizing scientific research during his very long presidency of JAE (1907–1932), and these efforts gave rise to the CSIC. The LIB became the Cajal Institute (IC) in 1932, and on November 24, 1939, the IC was incorporated to the CSIC, along with Cajal's legacy of reforms. Regarding the origin of CSIC (1939), no one can help but be aware of the fact that the father of this national organization was Santiago Ramón y Cajal.

Over its more than one hundred years of existence, the renowned scientists and professionals of the IC have contributed remarkable advances in neurobiology. Today, the IC is prepared to confront future challenges and to maintain its leading role in neurobiological research in Spain, always keeping in mind that the final destination of knowledge is the well-being of society.

Emilio Lora-Tamayo D'Ocón
President, CSIC
Madrid

The stunning drawings by Santiago Ramón y Cajal inspired *The Beautiful Brain* exhibition and book. They were brought to the Weisman Art Museum's attention by neuroscientists Eric A. Newman and Alfonso Araque. Araque had worked at the Cajal Institute in Madrid and knew of the treasure trove of original drawings that had rarely been seen outside Spain. Janet M. Dubinsky, also of our neuroscience department, joined the team to curate contemporary visualizations of the brain. Newman participated in curatorial decisions and also translated specialized information about each drawing into language accessible to a general audience. Three neuroscientists and I formed the unusual curatorial team; open-mindedness and acknowledgment of the expertise of each made it work.

We would like to extend our gratitude to the Spanish National Research Council (CSIC): Emilio Lora-Tamayo D'Ocón and José Ramon Urquijo Goitia, President and Vice President respectively; Maria del Carmen González Peñalver, Head of the Department of Infrastructure and Maintenance; José Antonio Ocaña, Head of Heritage Service; Pedro Ruiz de Arcaute, Head of Management Agreements; Ignacio Torres Alemán and Ricardo Martínez Murillo, Director and Vice Director of the Cajal Institute respectively. Gratitude is also due to Ramón Gil-Casares, Ambassador of Spain to the United States, and Carlos Gómez-Múgica Sanz, Ambassador of Spain to Canada. Special thanks to Ana Elorza, Science Coordinator and FECYT Representative at the Spanish Embassy in Washington, D.C., and Fernando Sánchez García for his excellent work scanning Cajal's drawings. Without their support and assistance, this book would not have been possible.

Lois Hendrickson, curator of the Wangensteen Historical Library of Biology and Medicine, helped select historical books as precedent for Cajal, assisted by Jacquelyn Gmiterko, museum intern. All museum staff worked with their usual creativity and efficiency but special acknowledgement to registrars, Annette Van Aken and Erin Bouchard. Eric Himmel of Abrams became a full partner; I am grateful for his knowledge and enthusiasm. Our mainstay was Curatorial Assistant Laura Moran, who made sure no detail was lost between here and Spain or any of the sites where the exhibit will be shown in the United States.

Lyndel King
Director and Chief Curator, Weisman Art Museum
University of Minnesota, Minneapolis

Flower still life, photograph by Cajal

Designer: Chin-Yee Lai
Production Manager: Anet Sirna-Bruder

Library of Congress Control Number: 2015955626

ISBN: 978-1-4197-2227-1

Copyright © 2017 Frederick R. Weisman Art Museum at the University of Minnesota

Photographs and drawings by Santiago Ramón y Cajal
© 2017 CSIC

Published in 2017 by Abrams, an imprint of ABRAMS. All rights reserved. No portion of this book may be reproduced, stored in a retrieval system, or transmitted in any form or by any means, mechanical, electronic, photocopying, recording, or otherwise, without written permission from the publisher.

Printed and bound in China
10 9 8 7 6 5 4 3 2 1

Abrams books are available at special discounts when purchased in quantity for premiums and promotions as well as fundraising or educational use. Special editions can also be created to specification. For details, contact specialsales@abramsbooks.com or the address below.

ABRAMS The Art of Books
115 West 18th Street, New York, NY 10011
abramsbooks.com

Credits:
Page 26: Courtesy of Frank Gehry
Page 29: Biblioteca Nacional de España, Madrid. © 2016 Artists Rights Society (ARS), New York/VEGAP, Madrid
Page 192: J. Livet, J. Sanes, and J. W. Lichtman, Harvard University. © Jeff Lichtman, reproduced with permission
Page 195: A. Wertz, S. Trenholm and B. Roska from the Friedrich Meischer Institute for Biomedical Research, Basel, Switzerland
197: D. Berger, N. Kasthuri, and J. W. Lichtman, Harvard University. © Jeff Lichtman, reproduced with permission
Page 198: S. Rizzoli, B. Rammner, University of Göttingen Medical School. From "Composition of isolated synaptic boutons reveals the amounts of vesicle trafficking proteins" by B. G. Wilhelm, S. Mandad, S. Truckenbrodt, K. Kröhnert, C. Schäfer, B. Rammner, S. J. Koo, G. A. Classe, M. Krauss, V. Haucke, H. Urlaub, S.O. Rizzoli, *Science,* 30 May 2014: 1023–1028. Reprinted with permission from AAAS
Page 99: C. Lenglet, K. Ugurbil, CMRR, UMN. Data from the Human Connectome Project, WU–Minn–Oxford Consortium (Principal Investigators: D. Van Essen and K. Ugurbil; 1U54MH091657) funded by the sixteen NIH Institutes and Centers that support the NIH Blueprint for Neuroscience Research; and by the McDonnell Center for Systems Neuroscience at Washington University
Page 201: Image and video generated by S. Smith, Oxford University, UK, from data acquired at the Center for Magnetic Resonance Research, University of Minnesota as part of the Human Connectome Project, WU–Minn–Oxford Consortium, and funded by the sixteen NIH Institutes and Centers that support the NIH Blueprint for Neuroscience Research